Universitext

For other titles in this series, go to
http://www.springer.com/series/223

Serge Alinhac

Hyperbolic Partial Differential Equations

Springer

Serge Alinhac
Université Paris-Sud XI
Département de Mathématiques
Orsay Cedex 91405
France
serge.alinhac@math.u-psud.fr

ISBN 978-0-387- 87822-5   e-ISBN 978-0-387-87823-2
DOI 10.1007/978-0-387-87823-2
Springer Dordrecht Heidelberg London New York

Library of Congress Control Number: 2009928133

Mathematics Subject Classification (2000): 35Lxx

Springer is part of Springer Science+Business Media (www.springer.com)

Contents

Introduction

The aim of this book is to present hyperbolic partial differential equations at an elementary level. In fact, the required mathematical background is only a third year university course on differential calculus for functions of several variables. No functional analysis knowledge is needed, nor any *distribution theory* (with the exception of shock waves mentioned below). All solutions appearing in the text are piecewise classical C^k solutions. Beyond the simplifications it allows, there are several reasons for this choice: First, we believe that all main features of hyperbolic partial differential equations (PDE) (well-posedness of the Cauchy problem, finite speed of propagation, domains of determination, energy inequalities, etc.) can be displayed in this context. We hope that this book itself will prove our belief. Second, all properties, solution formulas, and inequalities established here in the context of smooth functions can be readily extended to more general situations (solutions in Sobolev spaces or temperate distributions, etc.) by simple standard procedures of functional analysis or distribution theory, which are "external" to the theory of hyperbolic equations: The deep mathematical content of the theorems is already to be found in the statements and proofs of this book. The last reason is this: We do hope that many readers of this book will eventually do research in the field that seems to us the natural continuation of the subject: nonlinear hyperbolic systems (compressible fluids, general relativity theory, etc.). In this area, a large part of the work is devoted to prove global existence in time of classical solutions, in which case the whole work is about understanding the behavior and decay of smooth solutions.

There are of course many excellent books and textbooks partially or completely devoted to the subject of hyperbolic equations, some of which are quoted in the References at the end. But having discarded the books clearly too difficult to read for a first approach or that use abundantly distribution theory and Sobolev spaces, we found it somewhat hard to indicate references providing an easy introductory exposition of such subjects as, for instance, inequalities for variable coefficient equations, geometrical optics, etc. (Or, the references were scattered in many different books.)

The content of this book can be roughly divided into two parts. The first part includes all aspects of the theory having to do with vector fields and integral curves:

i) Cauchy problem for vector fields and (linear) method of characteristics (Chapter 1);

ii) Differential operators or systems in the plane, which reduce to systems of coupled vector fields (Chapter 2);

iii) Quasilinear scalar equations and eikonal equations, solved by nonlinear methods of characteristics, involving weaving by vector fields (Chapter 3).

We believe this part especially intuitive and *easy to visualize*: It is what makes hyperbolic PDE so attractive. Chapter 4 is a short introduction to conservation laws in one space dimension (shocks, simple waves, rarefaction waves, Riemann problem, etc.), which uses the language of vector fields and characteristics. This is the only place where the concept of solution "in the sense of distribution" is needed, but it is easy to understand in the special case of shock waves.

The second part describes the world of the wave equation and its perturbations for space dimensions two or three. Our treatment here, though completely elementary, emphasizes concepts proved useful by recent research developments: Lorentz fields and Klainerman inequality, weighted inequalities, conformal energy inequalities, etc. Following this orientation, we insisted more on inequalities than on explicit or approximate solutions. Chapter 5 presents the classical solution formula, along with the geometry of Lorentz fields, null frames, etc. In Chapter 6, we teach the reader how to prove an energy inequality, starting from the simplest case of a strip to proceed to inequalities in domains of determination; we also include an improvement of the standard inequality, Morawetz and KSS inequalities, and conformal inequality. Finally, Chapter 7 is devoted to variable coefficient equations or symmetric systems: We present the available inequalities with their amplification factors, Klainerman "energy method," and we touch upon geometrical optics and parametrics.

The natural readership for this book comprises senior or graduate students in mathematics interested in PDE; But the book can also be used by researchers of other fields of mathematics or sciences seeking to learn the basic facts about techniques they have heard of. The chapters are essentially independent, the language of vector fields or submanifolds, which is widely used throughout the book, being presented in two short Appendices.

In some chapters, Notes at the end explain the sources and references for further learning. At many places (and especially for energy inequalities in Chapters 6 and 7), instead of writing the proofs of the Theorems in the traditional formal way, we have presented them as "do it yourself" instructions with clearly identified steps. Finally, about 100 exercises are proposed, so that this book may be a useful textbook.

Chapter 1

Vector Fields and Integral Curves

Throughout the book we will use the notation $\mathbf{R}_x^n$ to denote the space $\mathbf{R}^n$ with variable x; similarly, $\mathbf{R}_{x,t}^2$ will denote the plane with coordinates (x,t), and so on.

1.1 First Definitions

In $\mathbf{R}^n$ we denote coordinates by $x = (x_1, \ldots, x_n)$. However, we will often take coordinates (x, y) or (x, t) in the plane. The usual scalar product of two vectors x and y is denoted by dot notation:

$$x \cdot y = \Sigma x_i y_i.$$

Definition 1.1. *A vector field defined on a domain $\Omega \subset \mathbf{R}^n$ is a function*

$$X : \Omega \to \mathbf{R}^n,\ X(x) = (X_1(x), \ldots, X_n(x)).$$

We will always assume (unless otherwise specified) that the components $X_i(x)$ are C^1 real functions in Ω.

Definition 1.2. *An integral curve of the vector field X is a C^1 function*

$$x : I \to \Omega \subset \mathbf{R}^n,\ x(t) = (x_1(t), \ldots, x_n(t))$$

defined on some interval I of $\mathbf{R}$ for which $x'(t) = X(x(t))$.

S. Alinhac, *Hyperbolic Partial Differential Equations*, Universitext,
DOI 10.1007/978-0-387-87823-2_1, © Springer Science+Business Media, LLC 2009

Geometrically, this means that X is tangent to the curve at every point. Thanks to the Cauchy–Lipschitz theorem (see Appendix, Theorem A.1), there is through each $x^0 \in \Omega$ a unique integral curve with $x(0) = x^0$.

Remark: A C^1 change of parameter $t = \phi(s)$ yields the new curve $y(s) = x(\phi(s))$, to which X is also tangent, since $y'(s) = (\phi'(s))X(y(s))$. Hence, the definition we have chosen corresponds to a specific choice of the parameter and avoids ambiguity in practice.

Example 1.3. If X is constant, the integral curves are straight lines parallel to X.

Example 1.4. In the plane, let $X(x, y) = (y, -x)$. The integral curves of X are defined by

$$x'(t) = y(t), \; y'(t) = -x(t).$$

The curve with starting point (x_0, y_0) given by

$$x(t) = x_0 \cos t + y_0 \sin t, \; y(t) = y_0 \cos t - x_0 \sin t,$$

is the circle centered at the origin through (x_0, y_0).

Example 1.5. In the plane, the integral curves of $X(x, y) = (x, 1)$ are defined by

$$x'(t) = x(t), \; y'(t) = 1.$$

The curve with starting point (x_0, y_0) is $x(t) = x_0 e^t$, $y(t) = y_0 + t$.

Example 1.6. In the plane, the integral curve of $X(x, t) = (x^2, 1)$ with starting point (x_0, t_0) is defined by

$$x'(s) = x^2(s), t'(s) = 1, x(0) = x_0, t(0) = t_0,$$

which gives $x(s) = x_0/(1 - x_0 s)$, $t(s) = t_0 + s$.

1.2   Flows

Let $X : \Omega \to \mathbf{R}^n$ be a vector field. To emphasize the way an integral curve of X depends on its starting point, we introduce the following definition.

Definition 1.7. *The flow of X is the function*

$$\Phi : \mathbf{R}_t \times \mathbf{R}^n_y \supset U \to \mathbf{R}^n$$

defined by

$$\partial_t \Phi(t, y) = X(\Phi(t, y)), \; \Phi(0, y) = y.$$

In other words, for fixed y, the function $t \mapsto \Phi(t, y)$ is just the integral curve of X starting from y at time $t = 0$. It is defined on an open maximal interval $I =]T_*(y), T^*(y)[$ (see Appendix, Theorem A.3), so that the domain of definition U of Φ is

$$U = \{(t, y) \in \mathbf{R}_t \times \Omega, t \in]T_*(y), T^*(y)[\}.$$

It can be shown that U is open (a nontrivial fact!) and $\Phi \in C^1(U)$. In general, using the definition and Taylor's formula, we obtain the approximation formula

$$\Phi(t, y) = y + tX(y) + (t^2/2)X'(y)X(y) + o(t^2), \ t \to 0,$$

where $X'(y) = D_y X$ is the $n \times n$ matrix that represents the differential of X.

Example 1.8. In the above Examples 1.4–1.6, the formulas given for $(x(t), y(t))$ define $\Phi(t, x_0, y_0)$. The domain U is the whole of $\mathbf{R}^3$, except in Example 1.6, where

$$U = V \times \mathbf{R}_y,$$

V being the open region of $\mathbf{R}_t \times \mathbf{R}_x$ between the two branches of the hyperbola $\{tx = 1\}$.

1.3 Directional Derivatives

Definition 1.9. *Fix x_0 and a in $\mathbf{R}^n$. For $f \in C^1$ in a neighborhood of x_0, the derivative of f at x_0 in the direction of a is*

$$\frac{d}{dt}[f(x_0 + at)](t = 0) = (a \cdot \nabla f)(x_0).$$

An analogous and *more flexible* definition is obtained by replacing the line through x_0 by a C^1 curve $\gamma = \{\gamma(t), t \in I\}$ with $\gamma(0) = x_0$. We define the *derivative of f at x_0 along γ* by

$$\frac{d}{dt}[f(\gamma(t))](t = 0) = \gamma'(0) \cdot \nabla f(x_0).$$

The important point here is that this derivative depends only on $\gamma'(0)$ and $\nabla f(x_0)$, and not on the actual curve γ in a neighborhood of x_0.

In the special case where γ is an integral curve of a field X, we obtain for all t the formula

$$\frac{d}{dt}[f(\gamma(t))] = (X \cdot \nabla f)(\gamma(t)).$$

The goal of this formula is to help us *visualize* the quantity $X \cdot \nabla f$, which occurs in many problems. This implies in particular the following theorem.

Theorem 1.10.   *Let X be a vector field in Ω and $f \in C^1(\Omega)$. Then $X \cdot \nabla f = 0$ in Ω if and only if f is constant along any integral curve of X in Ω.*

As a consequence, it is customary to identify the field X with the *operator* $X \cdot \nabla = \Sigma X_i \partial_i$ defined by

$$(X \cdot \nabla)u(x) = \Sigma X_i(x)(\partial_i u)(x).$$

We will say for instance "the field $a\partial_x + b\partial_y$," meaning $X = (a, b)$ in the plane, etc. We will write Xu (where X is considered an operator) for $X \cdot \nabla u$ (where X is considered a vector).

1.4   Level Surfaces

We give here two classical applications of derivatives along curves, in the context of submanifolds of $\mathbf{R}^n$ (see Appendix, Theorem A.11).

Proposition 1.11.a. *Let $f \in C^1(\mathbf{R}^n)$ be a real function with $\nabla f \neq 0$, and S be the submanifold of $\mathbf{R}^n$ defined by the single real equation*

$$S = \{x, f(x) = 0\}.$$

Then, for any $x_0 \in S$, $\nabla f(x_0)$ is orthogonal to $T_{x_0}S$.

Proof: In fact, for any C^1 curve $\gamma : t \mapsto \gamma(t)$ with $\gamma(0) = x_0$ drawn on S, f is zero along γ, hence its derivative along γ vanishes, that is, $\gamma'(0) \cdot \nabla f(x_0) = 0$. Since the vectors $\gamma'(0)$ span $T_{x_0}S$, the proposition is proved.   □

Proposition 1.11.b. *Let S be as in Proposition 1.11.a, and $g \in C^1(\mathbf{R}^n)$ be a real function such that its restriction to S has a minimum (or a maximum) at x_0 : then $\nabla g(x_0)$ is colinear to $\nabla f(x_0)$.*

Proof: The restriction of g to any C^1 curve γ on S has also a minimum (or a maximum) at x_0, hence its derivative $\gamma'(0) \cdot \nabla g(x_0)$ along γ is zero for $t = 0$. In other words, $\nabla g(x_0)$ is orthogonal to $T_{x_0}S$, hence colinear to $\nabla f(x_0)$.   □

1.5   Bracket of Two Fields

Definition 1.12. *Let $X(x) = \Sigma X_i(x)\partial_i$ and $Y(x) = \Sigma Y_j(x)\partial_j$ be two fields on $\Omega \subset \mathbf{R}^n$. The operator*

$$XY - YX$$

is a vector field, called the bracket of X and Y, and denoted by $[X, Y]$.

This follows from

$$\Sigma X_i \partial_i (\Sigma Y_j \partial_j u) = \Sigma\Sigma X_i Y_j \partial_{ij}^2 u + \Sigma X(Y_j)\partial_j u$$

and the fact that the second order terms cancel in the difference $(XY - YX)(u)$.

Example 1.13. In $\mathbf{R}^3$, for all i, j, $[\partial_i, \partial_j] = 0$. Also, $[\partial_1, \partial_2 + x_1\partial_3] = \partial_3$, and

$$[x_2\partial_3 - x_3\partial_2, x_3\partial_1 - x_1\partial_3] = -(x_1\partial_2 - x_2\partial_1).$$

1.6   Cauchy Problem and Method of Characteristics

Definition 1.14. *Let $\Sigma \subset \Omega \subset \mathbf{R}^n$ be a hypersurface and X a field on Ω. Given $f \in C^0(\Omega)$ and $u_0 \in C^1(\Sigma)$, the Cauchy problem for X with initial hypersurface Σ and data (u_0, f) is the problem of finding $u \in C^1(\Omega)$ with*

$$Xu = f, \ x \in \Sigma \Rightarrow u(x) = u_0(x).$$

If X is not tangent to Σ on a subdomain $\omega \subset \Sigma$, the union of all integral curves of X starting from points $x_0 \in \omega$ can be visualized as a "tube" $\mathcal{T}$ with base ω, called "domain of determination of ω." For $x^0 \in \mathcal{T}$, the integral curve starting from x^0 intersects Σ at a point $p(x^0)$; if $f \equiv 0$, the solution u of the Cauchy problem necessarily satisfies

$$u(x^0) = u_0(p(x^0)).$$

This is the **"method of characteristics."** It remains to prove, however, that p is a C^1 function, so that u is actually a solution of the Cauchy problem. We will discuss this in Sections 1.7 and 1.8. The method of characteristics also allows us to handle the non-homogeneous Cauchy problem ($f \not\equiv 0$), as will be seen on examples.

Let us first examine some examples in the plane with coordinates (x, t), and $\Sigma = \{t = 0\}$:

Example 1.15. The simplest case is the Cauchy problem

$$\partial_t u = 0, \ u(x, 0) = u_0(x).$$

The solution is $u(x, t) = u_0(x)$. The integral curves of $X = \partial_t$ are vertical lines $\{x = C\}$, and u is constant on each one of them. To solve the inhomogeneous equation $\partial_t u = f$, one writes

$$u(x, t) = u(x, 0) + \int_0^t f(x, s)ds = u_0(x) + \int_0^t f(x, s)ds.$$

We say that the function u is obtained from u_0 by integration of f along the integral curves of ∂_t.

Example 1.16. The next case is the **advection equation**

$$\partial_t u + a\partial_x u = f, \ u(x, 0) = u_0(x),$$

where a is a real constant. The integral curves of $X = (a, 1)$ are the lines $(x(t) = x_0 + at, t)$, and

$$\frac{d}{dt}(u(x_0 + at, t)) = (Xu)(x_0 + at, t) = f(x_0 + at, t).$$

Thus

$$u(x, t) = u_0(x - at) + \int_0^t f(x + a(s - t), s)ds.$$

For $f \equiv 0$, the equation is thought of representing a *propagation at speed* a, since the graph of $u(\cdot, t)$ is just the graph of u_0 translated by at.

Example 1.17. Consider now the Cauchy problem for the field of Example 1.5:

$$\partial_t u + x\partial_x u = f, \ u(x, 0) = u_0(x).$$

The domain of determination of Σ is here the whole of $\mathbf{R}^2$. For (x_0, t_0) given, the point p is

$$p(x_0, t_0) = (x_0 e^{-t_0}, 0)$$

and the solution u of the Cauchy problem for $f \equiv 0$ is then $u(x, t) = u_0(xe^{-t})$. In the inhomogeneous case, we write

$$\frac{d}{dt}(u(x_0 e^t, t)) = f(x_0 e^t, t),$$

which gives

$$u(x,t) = u_0(xe^{-t}) + \int_0^t f(xe^{s-t}, s)ds.$$

Example 1.18. In the case of Example 1.6, we have for $t \geq 0$ on the integral curve starting at $(x_0 < 0, 0)$,

$$-xt = -\frac{x_0 t}{1 - x_0 t} \leq 1.$$

It is left as an exercise (Exercise 5) to show that the domain of determination of Σ is only $\{(x,t), xt \geq -1\}$, and there the solution of the homogeneous equation is

$$u(x,t) = u_0\left(\frac{x}{1 + xt}\right).$$

Important Remark: Consider in the plane $\mathbf{R}_{x,t}^2$ the Cauchy problem

$$\partial_t u + a(x,t)\partial_x u = 0, \ u(x,0) = u_0(x)$$

for some *complex* function a ($\Im a \neq 0$): We claim that this Cauchy problem cannot be "well-posed." In the case where a is constant, two different kinds of arguments can be used to justify this claim: First, it is well-known (think of the case $a = i$ of the Cauchy–Riemann operator) that the solutions of the equation $\partial_t u + a\partial_x u = 0$ are analytic functions of (x,t), hence u_0 must also be real analytic; no solution exists for $u_0 \in C^\infty$ in general. The second argument is more subtle and due to Hadamard: suppose $a = \alpha + i\beta, \beta > 0$, and consider the functions

$$v_n(x,t) = e^{-\sqrt{n}}e^{in(x-at)}, \ n \in \mathbf{N},$$

which are solutions of the equation. Then $v_n(x,0)$ and all its derivatives are small in any C^k norm, while v_n is very big for $t = t_0 > 0$ and large n. No reasonable control of the solution v_n by its data $v_n(x,0)$ is to be expected.

In the more general case of a complex function $a(x,t)$, the first argument no longer works, but the second can be modified to show that, even if the solution u were to exist and be unique for all data u_0, it would not depend reasonably of u_0; it is this concept of "continuous dependence" (in a sense we should make, of course, more precise) which is at the heart of Hadamard's concept.

1.7   Stopping Time

We come back now to the smoothness of the function p defined at the beginning of Section 1.6. Let us assume that for some x^0 the integral curve of X starting at x^0 intersects Σ at time T_0 at the point $p_0 = \Phi(T_0, x^0)$ (Φ is the flow of X). For x close to x^0, we look for $T(x)$ close to T_0 such that the integral curve of X starting from x intersects Σ at time $T(x)$ (the "stopping time").

Proposition 1.19. *Assume that Σ is defined near p_0 as $\{f = 0\}$ for some $f \in C^1$, $\nabla f(p_0) \neq 0$. If $X(p_0)$ in not tangent to Σ, then $T \in C^1$ in a neighborhood of x^0.*

Proof: We want to solve for T the equation $f(\Phi(T, x)) = 0$ for x given close to x^0. This can be done using the implicit function theorem, provided

$$\partial_t[f(\Phi(t, x))](T_0, x^0) = (X \cdot \nabla f)(p_0) \neq 0,$$

that is, if X is not tangent to Σ at p_0 (which is an obvious necessary condition; See Exercise 4). The intersection point is then $p(x) = \Phi(T(x), x)$, and p also is a C^1 function. $\square$

1.8   Straightening Out of a Field

An alternative approach to that of the preceding section (Section 1.7) is contained in the following proposition.

Proposition 1.20. *Let X be a field in $\Omega \subset \mathbf{R}^n$, $0 \in \Omega$, with $X(0) \neq 0$. Then there exists a C^1 diffeomorphism Ψ from a neighborhood U of 0 onto a neighborhood V of 0, $\Psi(0) = 0$, such that the image by Ψ of the integral curves of X in U are parallel lines.*

Proof: To see this, let us assume $X_n(0) \neq 0$, and, since we are dealing only with integral curves, assume in fact $X_n \equiv 1$ close to the origin. Let Φ be the flow of X. Consider now the map

$$(x', t) \mapsto F(x', t) = \Phi(t, (x', 0)), \quad x' = (x_1, \ldots, x_{n-1}).$$

Since the differential $D_0 F$ is an upper triangular matrix with one on the diagonal, it is invertible, and the implicit function theorem shows that F is a local diffeomorphism taking $(0, 0)$ to itself; we will take then $\Psi = F^{-1}$. Using the notation of Section 1.7 with $\Sigma = \{x_n = 0\}$, we see that Ψ contains both the information about the stopping time and the intersection point, $\square$

$$\Psi(x) = (\tilde{p}(x), T(x)), \quad p(x) = (\tilde{p}(x), 0).$$

1.9 Propagation of Regularity

Proposition 1.21. *Let X be a C^∞ field on $\Omega \subset \mathbf{R}^n$. Assume given some function $u \in C^1(\Omega)$ such that*

$$Xu = f \in C^\infty(\Omega).$$

If $u \in C^k$ in a neighborhood of $x_0 \in \Omega$, then $u \in C^k$ in a neighborhood of all points of the integral curve of X starting from x_0.

Proof: The simplest way of proving this is perhaps to straighten out the field X: according to Proposition 1.20 and its proof, the diffeomorphism Ψ is in this case a C^∞ diffeomorphism, so it is enough to prove the proposition for $X \equiv \partial_n$: In this case, the explicit formula of Example 1.15 yields the result. $\qquad\Box$

1.10 Exercises

1. Let S^2 be the unit sphere in $\mathbf{R}^3$, and e be a tangent vector to it at m. Let f be a C^1 function on $\mathbf{R}^3$ vanishing on the sphere. Show that the derivative of f at m in the direction of e is zero.

2. In the plane with coordinates (x,y), let $\mathbf{R}_+ = \{(x,0), x \geq 0\}$ and $A = \mathbf{R}^2 - \mathbf{R}_+$. Find $f \in C^\infty(A)$, not independent of y, and satisfying $\partial_y f = 0$.

3. Let X be a nonvanishing field in $\mathbf{R}^n$ and $a \in C^1(\mathbf{R}^n)$ a complex function. Explain how the study of the equation $Xu + au = f$ can be (locally) reduced to the study of the equation $Xv = g$.

4. In the context of Section 1.7, give an explicit example where the stopping time, defined as $T(x^0)$ for the point x^0, is not necessarily defined near x^0.

5. Let $X = \partial_t + x^2 \partial_x$ be the field in the plane considered in Example 1.18. Show that the domain of determination of $\{t = 0\}$ for X is $\{(x,t), xt \geq -1\}$. Construct a solution u of $Xu = 0$ in the plane, vanishing on $\{t = 0\}$ but not identically zero.

6. Let $u : \mathbf{R}^2_{x,t} \supset \Omega \to \mathbf{R}$ be a C^1 function and consider the field $X = \partial_t + u\partial_x$. Show that the integral curves of X are straight lines if and only if u is solution of Burgers equation

$$\partial_t u + u\partial_x u = 0.$$

7. Let $u : \mathbf{R}^2_{x,t} \supset \Omega \to \mathbf{R}$ be a C^1 function solution of the Cauchy problem

$$\partial_t u + u \partial_x u = u^2, \; u(x,0) = u_0(x).$$

Compute the integral curve of the field $X = \partial_t + u \partial_x$ starting from $(x_0, 0) \in \Omega$.

8. Let $D = \{(x,t) \in \mathbf{R}^2_{x,t}, \; x \geq 0, \; t \geq 0, \; x + t \leq 1\}$. For which values of the real constant λ does the Cauchy problem

$$(\partial_t + \lambda \partial_x) u = 0, \; u(x,0) = u_0(x)$$

have a unique solution in D?

9. Let a and $\alpha \geq 0$ be real constants and consider the field in the plane $\mathbf{R}^2_{x,t}$

$$X = t \partial_t + a t^\alpha \partial_x, \; t \geq 0.$$

Discuss, according to α, the behavior of the integral curves of X in the upperhalf plane. For which values of α does the Cauchy problem $Xu = f$, $u(x,0) = u_0(x)$, have at most one solution u, $u \in C^1(\{t > 0\}) \cap C^0(\{t \geq 0\})$?

10. Let $X_1 = (-x, y), X_2 = (-x, y + x^2)$ be two fields in the plane $\mathbf{R}^2_{x,y}$. Compute the flows Φ_1 and Φ_2 of these fields. Verify for each the flow property

$$\Phi(t_2, \Phi(t_1, x)) = \Phi(t_1 + t_2, x).$$

Compute for each field a function $F_i(x,y)$ such that F_i is constant along each integral curve of X_i. Find a C^∞ diffeomorphism D of the plane such that the image by D of an integral curve of X_1 is an integral curve of X_2. Show that one can arrange to have also $F_2(D) = F_1$.

11. Let X be a C^1 field on $\mathbf{R}^n$. Assume that, for some x_0, the flow $\Phi(t, x_0)$ is defined for all $t \in \mathbf{R}_+$ and $\Phi(t, x_0) \to a$ as $t \to +\infty$. Show that $X(a) = 0$.

12. Let S be a hypersurface in $\mathbf{R}^n$, and X a field tangent to S. Show that an integral curve of X starting from $x_0 \in S$ remains on S (Hint: Near a given point, choose coordinates so that S is defined by $\{x_1 = 0\}$).

13. Let S be a hypersurface in $\mathbf{R}^n$ and X a field tangent to S. Show that if $f \in C^1$ vanishes on S, so does Xf. Now let Y be another field tangent to S. Show that the bracket $[X, Y]$ is tangent to S.

14.(a) Consider in $\mathbf{R}^3$ the two fields $X_1 = \partial_1 + 2x_1 \partial_3, X_2 = \partial_2 + 2 \partial_3$. Check that their bracket is zero and compute their flows Φ_1 and Φ_2. Show that, for all x and t,

$$g(t, x) \equiv \Phi_2(t, \Phi_1(t, x)) = \Phi_1(t, \Phi_2(t, x)) \equiv h(t, x).$$

(b) Consider now the two fields $X_1 = \partial_1, X_2 = \partial_2 + x_1 \partial_3$. Compute their bracket and their flows Φ_1 and Φ_2. Show that in this case, in contrast with the preceding case,

$$g(t, x) - h(t, x) = t^2(0, 0, 1).$$

(c) More generally, for any two C^1 fields X_1 and X_2, using the preceding notation, show the formulas

$$\partial_t g(0, x) = X_1(x) + X_2(x),$$
$$\partial_t^2 g(0, x) = (X_1' X_1 + X_2' X_2 + 2 X_2' X_1)(x).$$

Deduce from these formulas

$$g(t, x) - h(t, x) = t^2[X_1, X_2](x) + O(t^3).$$

15.(a) Let us consider again the two fields of exercise 14(a): Check that $g = x_1^2 + 2x_2 - x_3$ satisfies $X_1 g = 0, X_2 g = 0$. In contrast, consider a function $g \in C^2(\mathbf{R}^3)$ satisfying $X_1 g = X_2 g = 0$ for the two fields of Exercise 14(b). Show that g is constant.

(b) Let us consider two independent fields X_1 and X_2 in the plane. Define the functions α and β by

$$[X_1, X_2] = \alpha X_1 + \beta X_2.$$

Write down a necessary condition on $f_i \in C^1$ for a C^2 solution u to exist for the system
$$X_1 u = f_1, X_2 u = f_2.$$

Show that this condition is (locally) sufficient.

16. Let u be a real C^1 solution of the equation

$$a(x, y)\partial_x u + b(x, y)\partial_y u = -u$$

in the closed unit disc D of the plane. We assume here that a and b are given C^1 real coefficients, with

$$a(x, y)x + b(x, y)y > 0$$

on the unit circle. Show that $u \equiv 0$ (Hint: One can show that u cannot have a positive maximum).

17. Let X be a constant coefficients vector field $X = \partial_t + \Sigma a_i \partial_i$ in $\mathbf{R}_t \times \mathbf{R}_x^n$. Let $\omega \subset \Sigma = \{t = 0\}$ be a compact domain with C^∞ boundary, and $\mathcal{T}$ the tube formed by the integral curves of X starting from ω. Denote by $\mathcal{T}_s$ the set

$$\mathcal{T}_s = \{x, (x, s) \in \mathcal{T}\}.$$

For a given $u : \mathbf{R}_t \times \mathbf{R}_x^n \to \mathbf{C}$, we define $E_t(u)$, the "energy" of u at time t, by

$$E_t(u) = \frac{1}{2} \int_{\mathcal{T}_t} |u(x, t)|^2 dx.$$

(a) Show the identity $2(Xu)(u) = \partial_t(u^2) + \Sigma \partial_i(a_i u^2)$.

(b) Using (a) and Stokes formula (see Appendix, A.2), prove for all *real* $u \in C^1$ the "energy identity"

$$E_t(u) = E_0(u) + \int_{\mathcal{T} \cap \{0 \le s \le t\}} (Xu)u \, dx \, ds.$$

(c) Deduce from (b) the "energy inequality"

$$E_t(u)^{1/2} \le C[E_0(u)^{1/2} + \int_0^t \|(Xu)(\cdot, s)\|_{L^2(\mathcal{T}_s)} ds],$$

for some appropriate constant C.

(d) Extend this result to *complex*-valued functions u.

Chapter 2

Operators and Systems in the Plane

2.1 Operators in the Plane: First Definitions

We will work in the plane $\mathbf{R}^2$ with coordinates (x, t).

Definition 2.1. *A differential operator P of order $m \in \mathbf{N}$ is defined by*

$$(Pu)(x, t) = \Sigma_{k+l \leq m} a_{kl}(x, t) \partial_x^k \partial_t^l u(x, t).$$

Here, the coefficients a_{kl} are C^∞, given functions (to simplify). The operator

$$P_m = \Sigma_{k+l=m} a_{kl}(x, t) \partial_x^k \partial_t^l$$

is the "principal part" of P, the rest $P - P_m$ being the "lower order terms."

Definition 2.2. *Assume $a_{0m} \neq 0$. The Cauchy problem for the differential operator P with initial surface $\Sigma = \{t = 0\}$ and data $(u_0, \ldots, u_{m-1}, f)$ is the problem of finding $u \in C^m$ such that*

$$Pu = f, \ u(x, 0) = u_0(x), \ldots, (\partial_t^{m-1} u)(x, 0) = u_{m-1}(x).$$

Here $f \in C^0$ and the m functions $u_0, \ldots, u_{m-1}$ are given with $u_k \in C^{m-k}$.

We remark that if $u \in C^\infty(\mathbf{R} \times [0, T[)$ is a solution of the Cauchy problem, all traces $(\partial_t^k u)(x, 0)$ are known from the data; in fact, using the equation

S. Alinhac, *Hyperbolic Partial Differential Equations*, Universitext,
DOI 10.1007/978-0-387-87823-2_2, © Springer Science+Business Media, LLC 2009

for $t = 0$, we obtain $(\partial_t^m u)(x, 0)$ from

$$a_{0m}(x, 0)(\partial_t^m u)(x, 0) + \Sigma_{k+l \leq m, l < m} a_{kl}(x, 0)\partial_x^k u_l(x) = f(x, 0).$$

Differentiating any number of times with respect to x yields $\partial_t^m \partial_x^p u(x, 0)$. Differentiating the equation once with respect to t gives us $\partial_t^{m+1}(x, 0)$, and so on.

Definition 2.3. *For fixed* (x, t), *the roots of the polynomial equation in* τ

$$\Sigma_{k+l=m} a_{kl}(x, t)\tau^l = 0$$

are denoted by $-\lambda_1(x, t), \ldots, -\lambda_m(x, t)$, *and the* λ_i *are called the characteristic speeds of* P.

This terminology can be understood from Example 1.16. Note that this equation involves only the coefficients of the principal part of P.

Definition 2.4 (Hyperbolicity). *We say that* P *is hyperbolic in* $\Omega \subset \mathbf{R}^2$ *if all the characteristic speeds* λ_i *are real in* Ω. *We call it strictly hyperbolic if they are also distinct.*

In dealing with the Cauchy problem, we will *always* make the assumption that P is hyperbolic. If P is strictly hyperbolic, the functions λ_i are C^∞ by the implicit function theorem, and we will order them

$$\lambda_1(x, t) < \cdots < \lambda_m(x, t).$$

We can also write $P = a_{0m}\Pi(\partial_t + \lambda_i(x, t)\partial_x) + Q$, where Q is an operator of order $m - 1$. Hence the principal part of P is just a product of *real* vector fields (modulo lower order terms).

Example 2.5. The one-dimensional wave equation (also called "vibrating string" equation) is (c being a positive constant)

$$P = \partial_t^2 - c^2 \partial_x^2 = (\partial_t - c\partial_x)(\partial_t + c\partial_x) = (\partial_t + c\partial_x)(\partial_t - c\partial_x).$$

The operator P is associated to the quadratic form $\tau^2 - c^2\xi^2$, the level sets of which are hyperbola in the plane (ξ, τ). This explains the denomination "hyperbolic." More generally, for a and b real constants satisfying $a^2 - 4b > 0$, the operator

$$P = \partial_t^2 + a\partial_{xt}^2 + b\partial_x^2$$

is strictly hyperbolic.

Example 2.6. The operator $P = \partial_t^2 - x^2\partial_x^2 = (\partial_t + x\partial_x)(\partial_t - x\partial_x) + x\partial_x$ is hyperbolic, but not strictly hyperbolic for $x = 0$.

Example 2.7. The operator $P = \partial_t^2 - t^2\partial_x^2 = (\partial_t + t\partial_x)(\partial_t - t\partial_x) + \partial_x$ is hyperbolic, but not strictly hyperbolic for $t = 0$.

Example 2.8. The Tricomi operator $P = \partial_t^2 + t\partial_x^2$ is strictly hyperbolic for $t < 0$, hyperbolic but nonstrictly for $t = 0$, and not hyperbolic for $t > 0$.

2.2   Systems in the Plane: First Definitions

Definition 2.9. *A first order system is an operator of the form*

$$L = S(x,t)\partial_t + A(x,t)\partial_x + B(x,t),$$

where S, A, and B are C^∞ $N \times N$ matrices, and L acts on C^1 vectors in $\mathbf{C}^N$ by

$$LU = S\partial_t U + A\partial_x U + BU.$$

Definition 2.10.   *Assume S invertible.   The Cauchy problem for the system L with initial surface $\Sigma = \{t = 0\}$ and data (U_0, F) is the problem of finding $U \in C^1$ such that*

$$LU = F, \ U(x,0) = U_0(x),$$

where $F \in C^0$ and $U_0 \in C^1$ are given.

Definition 2.11 (Hyperbolicity). *The system L is hyperbolic if all the eigenvalues λ_i of $S^{-1}A$ are real. These eigenvalues are called the characteristic speeds of L. We call L strictly hyperbolic if the characteristic speeds are distinct. The system is symmetric hyperbolic if S and A are hermitian and S is positive definite.*

The importance of symmetry is not obvious and is explained in Exercise 13. See also Chapter 7, Section 7.3, where the notion is discussed in detail.

If L is strictly hyperbolic, the eigenvalues $\lambda_i(x,t)$ are C^∞, since they are simple roots of a polynomial (the characteristic polynomial) with C^∞ coefficients. Assume that, in the domain D where we work, we can choose a basis of smooth eigenvectors $(r_1(x,t), \ldots, r_N(x,t))$ of $S^{-1}A$. Then

$$P(x,t)^{-1}S^{-1}(x,t)A(x,t)P(x,t) = \Lambda(x,t),$$

where P has the eigenvectors r_i as its columns, and Λ is diagonal. Setting $U = PV$, we find that the Cauchy problem

$$LU = F, \ U(x,0) = U_0(x)$$

is equivalent to the Cauchy problem

$$\partial_t V + \Lambda \partial_x V + CV = G, \ V(x,0) = V_0(x) = P^{-1}(x,0)U_0(x),$$

with

$$C = P^{-1}\partial_t P + \Lambda P^{-1}\partial_x P + P^{-1}S^{-1}BP, \ G = P^{-1}S^{-1}F.$$

The principal part of the system is now diagonal, the functions V_i satisfying the N equations

$$(\partial_t + \lambda_i(x,t)\partial_x)V_i(x,t) + \Sigma C_{ij}(x,t)V_j(x,t) = G_i(x,t), \ i = 1, \dots, N.$$

We think of this new system as *scalar equations coupled* through the coefficients C_{ij}. If L has constant coefficients S and A and is homogeneous (that is, $B \equiv 0$), then $C \equiv 0$ and we just have a collection of N scalar equations, which can be solved as explained in Chapter 1.

Important Remark: We explained in Chapter 1 why the Cauchy problem for a nonreal field could not be well-posed in the sense of Hadamard. Since a system with different speeds λ_k can be diagonalized, it follows that hyperbolicity is a necessary condition for the Cauchy problem for the system L to be well-posed.

2.3   Reducing an Operator to a System

Just like one does for ordinary differential equations, one can reduce scalar operators of order m to $m \times m$ first order systems. Assume that the operator P contains no terms of order less than $m - 1$.

- If u is a C^m solution of the Cauchy problem

$$Pu = f, \ u(x,0) = u_0(x), \dots, (\partial_t^{m-1}u)(x,0) = u_{m-1}(x),$$

we introduce as new unknowns the m functions

$$U_0 = \partial_x^{m-1}u, \dots, U_k = \partial_t^k \partial_x^{m-1-k}u, \dots, U_{m-1} = \partial_t^{m-1}u.$$

Then U is a C^1 solution of the Cauchy problem

$$\partial_t U_0 = \partial_x U_1, \dots, \partial_t U_{m-2} = \partial_x U_{m-1},$$
$$\partial_t U_{m-1} = -(a_{0m})^{-1}\Sigma_{k\geq 1}a_{kl}\partial_x U_l + (a_{0m})^{-1}f,$$
$$U_0(x,0) = \partial_x^{m-1}u_0(x), \dots, U_{m-1}(x,0) = u_{m-1}(x).$$

• Conversely, if U is a C^1 solution of the above Cauchy problem, we define u by

$$\partial_t^m u = \partial_t U_{m-1}, \ u(x,0) = u_0(x), \ldots, \ (\partial_t^{m-1}u)(x,0) = u_{m-1}(x).$$

Then we obtain successively

$$\partial_t^{m-1}u = U_{m-1}, \ \partial_x\partial_t^{m-2}u = U_{m-2}, \ldots, \ \partial_x^{m-1}u = U_0,$$

and u turns out to be a C^m solution of the Cauchy problem for P. Just as we did in Section 2.3, we emphasize the fact that, since an operator can be reduced to a system with the same characteristic speeds (see Exercise 9), these speeds must be real in order for the Cauchy problem to be well-posed.

Example 2.12. In the case $m = 2$, $U_0 = \partial_x u$, $U_1 = \partial_t u$, we obtain from the wave equation $P = \partial_t^2 - c^2\partial_x^2$ the system

$$\partial_t U_0 = \partial_x U_1, \ \partial_t U_1 = c^2\partial_x U_0 + f.$$

If we modify the procedure slightly by setting

$$U_0 = c\partial_x u, \ U_1 = \partial_t u,$$

we obtain a *symmetric* system. We can even try right away

$$U_0 = \partial_t u + c\partial_x u, \ U_1 = \partial_t u - c\partial_x u,$$

and obtain a *diagonal* system. We note that U_0 and U_1 are just then the factors of P.

Example 2.13. For $P = \partial_t^2 - x^2\partial_x^2$ of Example 2.6 above, we can try the same approach, setting

$$U_0 = x\partial_x u, \ U_1 = \partial_t u.$$

Then

$$\partial_t U_0 = x\partial_x U_1, \partial_t U_1 = x\partial_x U_0 - U_0 + f,$$

and again we obtain a *symmetric* system.

Example 2.14. If we try the same procedure for $P = \partial_t^2 - t^2\partial_x^2$, setting

$$U_0 = t\partial_x u, U_1 = \partial_t u,$$

we obtain now the system

$$\partial_t U_0 = t\partial_x U_1 + \frac{U_0}{t}, \ \partial_t U_1 = t\partial_x U_0 + f,$$

which is *singular* on $\{t = 0\}$. The difference with Example 2.13 is not just a consequence of our awkwardness: It reflects a true difference in the behavior of the solutions of the Cauchy problems.

Example 2.15. For the Tricomi operator we use $U_0 = \partial_x u$, $U_1 = \partial_t u$; To get a nice system, we multiply the first line by $-t$ and obtain the symmetric system

$$-t\partial_t U_0 + t\partial_x U_1 = 0, \ \partial_t U_1 + t\partial_x U_0 = f.$$

Note that the system is symmetric hyperbolic exactly when $t < 0$.

If the operator P has terms of order less than $m - 1$, one can try to express them in terms of the new unknowns. For instance, if $m = 2$, $u(x,t) = u_0(x) + \int_0^t U_1(x,s)ds$, etc. The obtained system will not be strictly speaking a first order system, but the additional (integral) terms can be handled as zero order terms and cause no trouble.

For the operator in Example 2.13, if one chooses U_0 and U_1 as indicated in order to obtain a symmetric system, it will not be possible to express smoothly a lower order term such as $a(x,t)\partial_x u$ with the help of U, unless $a(0,t) = 0$. In fact, it can be shown that the well-posedness of the Cauchy problem for $P = \partial_t^2 - x^2\partial_x^2 + a(x,t)\partial_x$ requires precisely this condition. Thus, turning a nonstricly hyperbolic operator into a hyperbolic symmetric system is a subtle issue, one that requires sometimes additional conditions on the lower order terms, called "Levy conditions."

2.4   Gronwall Lemma

The following elementary lemma will be useful here and later on.

Lemma 2.16 (Gronwall Lemma). *Let $A, \phi \in C^0([0, T[)$ such that, for $0 \le t < T$,*

$$\phi(t) \le C + \int_0^t A(s)\phi(s)ds.$$

Assume that $A \ge 0$. Then $\phi(t) \le C \exp(\int_0^t A(s)ds)$.

The proof is left as Exercise 3.

2.5   Domains of Determination I (A priori Estimate)

Definition 2.17. *For a hyperbolic operator* P, *the field* $\partial_t + \lambda_i \partial_x$ *is called the* i-*characteristic field, and its integral curves are called* i-*characteristics of* P. *The same definition holds for first order systems.*

Note that we have shown that P is equal to the product of its characteristic fields (up to lower order terms) and that a system can be reduced to the diagonal system of its characteristic fields (modulo zero order coupling terms).

Definition 2.18. *A closed domain* $D \subset \mathbf{R}_x \times [0, \infty[$ *with base*

$$\omega = D \cap \{t = 0\}$$

is a domain of determination of ω *for an operator* P *(or a system* L*) if for any* $m = (x_0, t_0) \in D$, *and all* i, *the backward* i-*characteristic (that is, for* $t \leq t_0$*) drawn from* m *reaches* ω *while remaining in* D.

Example 2.19. Consider the wave equation, and take $\omega = [a, b]$ on the x-axis. A triangle D bounded by a line through $(a, 0)$ (with positive slope) and a line through $(b, 0)$ (with negative slope) is a domain of determination if the lines have slopes respectively less than c and greater than $-c$. The biggest possible D is bounded by lines with slopes c and $-c$, respectively. More generally, as a consequence of the usual comparison theorem for solutions of ordinary differential equations (see Appendix, Theorem A.7), we have the following theorem.

Theorem 2.20. *For a strictly hyperbolic operator or system, the biggest domain of determination* D *with base* $\omega = [a, b]$ *on the* x-*axis is the curved triangle bounded by the* x-*axis, the fastest characteristic (corresponding to* λ_m*) from* $(a, 0)$, *and the slowest characteristic (corresponding to* λ_1*) from* $(b, 0)$.

For a domain of determination D, we will denote by $p_i(m)$ the point where the backward i-characteristic $\gamma_i(m) = \{(x_i(t, m), t)\}$ drawn from m meets ω. We can now prove the following *a priori estimate*.

Theorem 2.21. *Let* D *be a compact domain of determination with base* ω *on the* x-*axis for a first order strictly hyperbolic system* L. *Set* $D_t = \{x, (x, t) \in D\}$. *Then there exists a constant* C *such that, for any* $U \in C^1(\bar{D})$,

$$\max_{0 \leq s \leq t} ||U(\cdot, s)||_{L^\infty(D_s)} \leq C\{||U_0||_{L^\infty(\omega)} + \int_0^t ||(LU)(\cdot, s)||_{L^\infty(D_s)} ds\}.$$

Proof: As explained in Section 1.6, we reduce the Cauchy problem $LU = F, U(x,0) = U_0(x)$ to the problem

$$\partial_t V + \Lambda \partial_x V + CV = G, V(x,0) = V_0(x).$$

Integrating the equation for V_i along the i-characteristic between 0 and t, we obtain

$$V_i(m) = (V_0)_i(p_i(m)) + \int_0^t [G_i - (CV)_i](x_i(s,m),s)ds.$$

We fix t and take the sup norm in x to get, for some numerical constant C_1,

$$||V_i(\cdot,t)||_{L^\infty(D_t)} \le ||V_0||_{L^\infty(\omega)} + C_1 \int_0^t \{||F(\cdot,s)||_{L^\infty(D_s)} + ||V(\cdot,s)||_{L^\infty(D_s)}\}ds.$$

We set now $\phi(t) = \max_{0 \le s \le t} ||V(\cdot,s)||_{L^\infty(D_s)}$. Summing the above inequalities over i, we obtain for $0 \le t' \le t \le T$ (with another constant C_2)

$$||V(\cdot,t')||_{L^\infty(D_{t'})} \le C_2 ||V_0||_{L^\infty(\omega)} + C_2 \int_0^T ||F(\cdot,s)||_{L^\infty(D_s)} ds$$

$$+ C_2 \int_0^t ||V(\cdot,s)||_{L^\infty(D_s)} ds.$$

Taking the supremum in t' we get for $t \le T$

$$\phi(t) \le A + C_2 \int_0^t \phi(s)ds, A = C_2 ||V_0||_{L^\infty(\omega)} + C_2 \int_0^T ||F(\cdot,s)||_{L^\infty(D_s)} ds.$$

Using the Gronwall lemma, we finally get $\phi(t) \le C_3 A$, which is the desired result. $\square$

In particular, the theorem implies the uniqueness of a possible solution to the Cauchy problem in D. From the proof of the theorem, we see that it can be extended to a noncompact domain (for instance, a strip $\{0 \le t \le T\}$), provided the appropriate obvious assumptions on the coefficients of L have been made. Such a theorem is called an **a priori estimate**, since it applies to any U.

2.6   Domains of Determination II (Existence)

We prove now an existence theorem in a domain of determination D, chosen as in Section 2.5.

Theorem 2.22. *Let D be a compact domain of determination with base ω on the x-axis for a first order strictly hyperbolic system L. Let $F \in C^1(D)$ and $U_0 \in C^1(\omega)$. Then there exists a unique solution $U \in C^1(D)$ of the Cauchy problem*

$$LU = F, \ U(x, 0) = U_0(x).$$

Proof: *Step 1.* We resume the notation of the proof of Theorem 2.21. We first prove that the system on V, written in integral form

$$V_i(m) = (V_0)_i(p_i(m)) + \int_0^t [G_i - (CV)_i](x_i(s, m), s)ds,$$

has a C^0 solution in D. To this aim, we define a sequence $V^n \in C^0(D)$ by

$$V_i^{n+1}(m) = (V_0)_i(p_i(m)) + \int_0^t [G_i - (CV^n)_i](x_i(s, m), s)ds, \ V^0 = 0.$$

Introducing $\delta^n(t) = \|V^{n+1}(\cdot, t) - V^n(\cdot, t)\|_{L^\infty(D_t)}$, we obtain by subtracting the equations for $n + 1$ and n and taking the supremum for fixed t as before,

$$\delta^n(t) \le C_1 \int_0^t \delta^{n-1}(s)ds.$$

We claim now that for some constants c_0 and c_1, we have for all n, $\delta^n(t) \le c_0 c_1^n t^n / n!$. For $n = 0$, this is certainly true for c_0 big enough, which we now fix accordingly. Assume that this is true for n: then we get from the above inequality and the induction hypothesis

$$\delta^{n+1}(t) \le C_1 \int_0^t c_0 c_1^n \frac{s^n}{n!} ds = C_1 c_0 c_1^n \frac{t^{n+1}}{(n+1)!}.$$

This shows that the claim is true if $c_1 \ge C_1$. If $t \le T$ in D, we obtain then

$$\|V^{n+1} - V^n\|_{L^\infty(D)} \le c_0 \frac{(c_1 T)^n}{n!},$$

which is the general term of a convergent series. Hence, V^n converges uniformly in D to some $V \in C^0(D)$, which is a solution of the system on V written in integral form.

Step 2. However, this does not imply that V is C^1 and satisfies the differential system! To handle this difficulty, we set $W^n = \partial_x V^n$, which is allowed since in fact V^n belongs to $C^1(D)$ if F and U_0 do. Differentiating with respect to x the integral expression of V^{n+1}, we obtain

$$W_i^{n+1}(m) = \partial_x[(V_0)_i(p_i(m)) + \int_0^t G_i(x_i(s,m),s)ds] - \int_0^t [(\partial_x C)V^n$$
$$+ CW^n]_i(x_i(s,m),s)(\partial_x x_i(s,m))ds.$$

Just as before, we prove that V^n and W^n converge uniformly in D to continuous functions V and W. This implies that V admits a continuous partial derivative $\partial_x V = W$. Since

$$\partial_t V^{n+1} + \Lambda \partial_x V^{n+1} + CV^n = G,$$

$\partial_t V^n$ also converges uniformly to a continuous function. Hence V admits continuous partial derivatives and is in C^1. We can then differentiate the system in integral form satisfied by V to recover the original system, and this finishes the proof.   □

2.7   Exercises

1.(a) Consider in the plane $\mathbf{R}^2_{x,t}$ the wave operator $P = \partial_t^2 - \partial_x^2$. Prove that any C^2 function u of the form $u(x,t) = \phi(x+t)$ or $u(x,t) = \psi(x-t)$ satisfies $Pu = 0$. Deduce from this an explicit formula for the solution u of the homogeneous Cauchy problem in a domain

$$D = \{(x,t), t \geq 0, t + |x| \leq a\}.$$

(b) Find explicitly the solution of the Cauchy problem $Pu = f$ in D with zero Cauchy data on $\{t = 0\}$.

2. Let D be the unit closed disc in the plane with coordinates (x,y), and ∂D the unit circle. What are all the C^2 solutions of the equation $\partial^2_{xy}u = 0$ in $\mathbf{R}^2$? in D? Show that the boundary value problem in D

$$\partial^2_{xy}u = f, u|\partial D = u_0$$

does not have a unique solution. If we impose the stronger boundary conditions $u = \nabla u = 0$ on ∂D, show that the corresponding boundary value problem in D has at most one solution. Write down necessary conditions on f for such a solution to exist.

3. Prove the Gronwall lemma (Section 2.4)
(Hint: Set $\psi(t) = C + \int_0^t A(s)\phi(s)ds$, and solve the differential inequality on ψ).

4. We consider a C^2 real solution u of the wave equation

$$Pu = (\partial_t^2 - \partial_x^2)u = 0$$

in the cylinder $\mathcal{C} = \{(x,t), t \geq 0, a \leq x \leq b\} \subset \mathbf{R}_{x,t}^2$. Assume that u satisfies the boundary conditions

$$u(a,t) = 0, \quad (\partial_t u + \partial_x u)(b,t) = 0.$$

(a) Define the energy of u at time t by

$$E(t) = \frac{1}{2} \int_a^b [(\partial_t u)^2 + (\partial_x u)^2](x,t)dx.$$

By computing $\int_{\mathcal{C} \cap \{0 \leq t \leq T\}} (Pu)(\partial_t u)dxdt$, show

$$E(T) - E(0) = -\int_0^T (\partial_t u)^2(b,t)dt.$$

The energy is said to "dissipate" along the boundary $\{x = b\}$.

(b) Show that for $t \geq 2(b - a)$, $u \equiv 0$ (so much energy dissipated that there is nothing left!).

5. Prove an a priori estimate analogous to that of Theorem 2.21 for a second order strictly hyperbolic operator P.

6. Prove an existence theorem analogous to that of Theorem 2.22 for a second order strictly hyperbolic operator P.

7. Let P be a strictly hyperbolic operator of order two in $\mathbf{R}_{x,t}^2$, and $u \in C^2(\mathbf{R}_x \times \mathbf{R}_t^+)$ be a solution of $Pu = 0$. Assume that the Cauchy data of u vanish outside $[a,b]$. Let $x = x_1(t)$ be the 1-characteristic of P through $(a,0)$, and $x = x_2(t)$ the 2-characteristic through $(b,0)$. Prove that the support of u is contained in the set

$$\{(x,t), t \geq 0, x_1(t) \leq x \leq x_2(t)\}.$$

8. Consider a strictly hyperbolic homogeneous operator P with constant coefficients. Show that if D is not a domain of determination of its base $[a,b]$ for P, no uniqueness can hold for the Cauchy problem in D.

9. Prove that when an operator P is reduced to a first order system L as in Section 2.3 the characteristic speeds are the same for P and L.

10. Let A be a real square matrix. Show that if there exists a hermitian positive definite S such that SA is hermitian, then the eigenvalues of A are

real. Conversely, if all eigenvalues of A are real and distinct, there exists such an S. Explain why this is relevant for hyperbolic systems.

11.(a) Let P be the wave operator with real coefficient $c \in C^1(\mathbf{R}^2)$

$$P = \partial_t^2 - c^2(x, t)\partial_x^2, \ 1/2 \le c \le 2.$$

Prove for all $u \in C^2(\mathbf{R}^2)$ the identity

$$(Pu)(\partial_t u) = \frac{1}{2}\partial_t[c^2(\partial_x u)^2 + (\partial_t u)^2] - \partial_x[c^2(\partial_x u)(\partial_t u)]$$
$$+ 2c(\partial_x c)(\partial_x u)(\partial_t u) - c(\partial_t c)(\partial_x u)^2.$$

(b) Assume that in the strip $S_T = \{0 \le t \le T\}$ for some constant C,

$$|\partial_x c| + |\partial_t c| \le C.$$

Assume for simplicity that u is real and that $u(\cdot, t)$ has compact support for all t. Using the formula of (a) to compute $\int_{S_t}(Pu)(\partial_t u)dxds$, prove for $t \le T$ the inequality

$$E(t) \le E(0) + C_1 \int_0^t E(s)ds + C_1 \int_0^t \|f(\cdot, s)\|_{L^2}E^{1/2}(s)ds,$$

where $Pu = f$ and $E(t) = (1/2) \int [c^2(\partial_x u)^2 + (\partial_t u)^2]dx$. Proceed then as in Exercise 17 of Chapter 1, using the Gronwall lemma, to establish the a priori L^2 inequality

$$\max_{0 \le s \le t} E^{1/2}(s) \le C_2 E^{1/2}(0) + C_2 \int_0^t \|f(\cdot, s)\|_{L^2}ds, \ t \le T.$$

Such an a priori inequality in L^2 norm is called an "energy inequality."

12. We keep the notation of Exercise 11. Let D be a compact domain of determination for P, and set $D_T = \{(x, t) \in D, 0 \le t \le T\}$. On the nonhorizontal part Λ of the boundary of D_T, we denote the components of the unit outgoing normal by $(n_x, n_t > 0)$. Proceeding as in Exercise 11, prove the a priori inequality

$$E(T) + \int_\Lambda (n_t^2 - c^2 n_x^2)\frac{(\partial_t u)^2}{2n_t}d\sigma \le E(0) + C_1 \int_0^T E(t)dt$$
$$+ C_1 \int_0^T \|f(\cdot, t)\|_{L^2}E^{1/2}(t)dt,$$

where $d\sigma$ is the length element on Λ and E is now defined by an integration on $D \cap \{t = T\}$. If $|n_x| \leq n_t/c$ on Λ, this yields exactly the same energy inequality as in Exercise 11. Show that this condition on ∂D is always satisfied for a domain of determination (this is a remarkable fact, since it shows that the method of proof does not require more assumptions than what is known to be necessary anyway).

13.(a) Let $L = S\partial_t + A\partial_x + B$ be a symmetric hyperbolic system, where we take for simplicity S and A to be real. Prove, for all real $U \in C^1(\mathbf{R}^2)$, the identity

$$2\,^tULU = \partial_t(^tUSU) + \partial_x(^tUAU) - \,^tU(\partial_t S + \partial_x A - 2B)U.$$

Give appropriate conditions on the coefficients of L in a strip $S_T = \{0 \leq t \leq T\}$ to obtain, as in Exercise 11, the energy inequality

$$\max_{0 \leq s \leq t} \|U(\cdot, s)\|_{L^2} \leq C_1 \|U_0\|_{L^2} + C_1 \int_0^t \|f(\cdot, s)\|_{L^2} ds.$$

(b) We keep the notation of Exercise 12 and set $E(t) = \int_{(x,t) \in D} |U(x,t)|^2 dx$. Prove the inequality

$$\|U(\cdot, T)\|_{L^2}^2 + \int_\Lambda \,^tU(n_t S + n_x A)U d\sigma \leq C_2 \|U_0\|_{L^2}^2 + C_2 \int_0^T E(t) dt$$

$$+ C_2 \int_0^T \|f(\cdot, t)\|_{L^2} E^{1/2}(t) dt.$$

Show that the conditions

$$n_x > 0 \Rightarrow n_t + \lambda_1 n_x \geq 0, \quad n_x < 0 \Rightarrow n_t + \lambda_N n_x \geq 0$$

imply that the matrix $n_t S + n_x A$ is nonnegative. Prove then an energy inequality analogous to that of (a). Are these conditions always satisfied for a domain of determination?

Chapter 3

Nonlinear First Order Equations

3.1 Quasilinear Scalar Equations

We consider in $\mathbf{R}_x^n$ the Cauchy problem with data u_0 given on the initial surface $\Sigma_0 = \{x_n = 0\}$ for the quasilinear scalar equation

$$\Sigma a_i(x, u)\partial_i u(x) = b(x, u), \ u(x', 0) = u_0(x'), \ x' = (x_1, \ldots, x_{n-1}).$$

The coefficients $a = (a_1, \ldots, a_n)$ and b are given real C^∞ functions on $\mathbf{R}_x^n \times \mathbf{R}_u$, and $u_0 : \mathbf{R}^{n-1} \to \mathbf{R}$ is a given C^1 function. We look for a C^1 real solution

$$u : \mathbf{R}_x^n \supset \Omega \to \mathbf{R}.$$

First we tranform this Cauchy problem into a purely *geometric* problem. To a real function $u \in C^1(\Omega)$ we associate its graph in $\Omega \times \mathbf{R}_z$,

$$S = \{(x, z), x \in \Omega, \ z = u(x)\}.$$

The **method of characteristics** is based upon the following observation:

Observation 3.1. *The function u is a solution of the equation in Ω if and only if the field*

$$V(x, z) = (a_1(x, z), \ldots, a_n(x, z), \ b(x, z)) \in \mathbf{R}^{n+1}$$

is tangent to S.

S. Alinhac, *Hyperbolic Partial Differential Equations*, Universitext,
DOI 10.1007/978-0-387-87823-2_3, © Springer Science+Business Media, LLC 2009

Note that the field V is given with the coefficients a and b and does not depend on u. Since a normal to S is

$$N = \nabla(u(x) - z) = (\partial_1 u, \dots, \partial_n u, -1),$$

and, at a point of S,

$$(N \cdot V)(x, u(x)) = \Sigma \partial_i u(x) a_i(x, u(x)) - b(x, u(x)),$$

the observation is proved.   □

Thus the Cauchy problem for u is equivalent to the following **geometric problem**: find an n-submanifold S in $\mathbf{R}^{n+1}$ such that

i) S is the graph of some function;

ii) S contains the $(n-1)$-submanifold $\Sigma = \{(x, z), x \in \Sigma_0, z = u_0(x)\}$;

iii) V is tangent to S.

Since V is tangent to S, the integral curves of V starting from points of Σ belong to S (see Exercise 1.12). Hence we take for S the union of the integral curves of V starting from Σ. We call this construction "weaving" (see Appendix, Theorem A.15).

We have however to be careful about the following questions:

1. Is S a submanifold?

2. Is S the graph of a C^1 function?

The answers to these questions can be only local in general.

1. Given a point $m_0 = (x_0, u_0(x_0)) \in \Sigma$, S is an n-submanifold in a neighborhood of m_0 if V is not tangent to Σ at m_0 (see "weaving," Appendix, Theorem A.15).

2. For S to be the graph of a smooth function near m_0, it is enough to have $a_n(m_0) \neq 0$. In fact, $T_{m_0} S$ is spanned by $T_{m_0} \Sigma$ and $V(m_0)$ and cannot contain a vertical vector W unless W is already in $T_{m_0} \Sigma$, which is impossible. Now, if $f = 0$ is an equation of S (with $\nabla f \neq 0$), the condition about $T_{m_0} S$ implies $\partial_z f \neq 0$. Then, using the implicit function theorem, we can solve in C^1 the equation $f = 0$ in z, that is, S is the graph of a C^1 function (see Appendix, Theorem A.14 for a more general statement).

To summarize, we have proved the following theorem.

Theorem 3.2. *Let $x_0 \in \Sigma_0$ and assume $a_n(x_0, u_0(x_0)) \neq 0$. Then there exists a unique solution $u \in C^1$ of the Cauchy problem in a neighborhood of x_0, and its graph is the union of the integral curves of V starting from Σ.*

Observe also that the projections on $\mathbf{R}_x^n$ of the integral curves of V are just integral curves of the field $x \mapsto a(x, u(x))$.

Though the statement of the above theorem is only local, the method sometimes yields global results, as shown in examples.

Example 3.3. Suppose we deal with a *linear* equation, i.e., a and b do not depend on u. The integral curves of V are now defined by

$$x'(s) = a(x(s)), \quad z'(s) = b(x(s)).$$

The x-part corresponds to integral curves of a; the z-part corresponds to integrating b along the characteristic curves. Thus we are back to the (linear) method of characteristics explained in Chapter 1.

Example 3.4. In the plane with coordinates (x, t), consider the Cauchy problem for **Burgers equation**

$$\partial_t u + u \partial_x u = 0, \quad u(x, 0) = u_0(x).$$

In this case, $V(x, t, z) = (z, 1, 0)$, and the integral curve of V starting from $(x_0, 0, z_0)$ satisfies

$$x'(s) = z(s), \quad t'(s) = 1, \quad z'(s) = 0,$$
$$x(s) = x_0 + s z_0, \quad t(s) = s, \quad z(s) = z_0.$$

Thus S is the image of

$$F : (y, s) \mapsto F(y, s) = (y + s u_0(y), s, u_0(y)).$$

Though S is a union of horizontal lines, it is not necessarily everywhere the graph of a smooth function! To see this, observe that at a point $m = (y + s u_0(y), s, u_0(y)) \in S$, $T_m S$ is spanned by

$$\partial_y F = (1 + s u_0'(y), 0, u_0'(y)), \quad \partial_s F = (u_0(y), 1, 0).$$

There exists a nontrivial vertical linear combination of these two vectors if and only if

$$1 + s u_0'(y) = 0.$$

If u_0' is not everywhere nonnegative, such points always exist (see Exercises 1–7 for a more precise discussion).

The method of characteristics is an **effective way** of finding solutions. In the case of Burgers equation, for instance, finding u reduces to solving in y, for given (x, t), the equation

$$y + tu_0(y) = x,$$

and then take $u(x, t) = u_0(y)$. For instance, if we take $u_0(x) = x^2$, we get for (x, t) close to zero

$$y = \frac{-1 + (1 + 4xt)^{1/2}}{2t}, \quad u(x, t) = y^2.$$

In practice, one often proceeds as follows: Suppose the solution u exists in a domain containing the x-axis. The characteristic (that is, the integral curve of $\partial_t + u\partial_x$) starting from $(x_0, 0)$ is then

$$t \mapsto (x = x_0 + tu_0(x_0), t)$$

and $u(x, t) = u_0(x_0)$ on it. In other words,

$$u(x, t) = u_0(x - tu(x, t)),$$

and one can solve this implicit equation for u to get the solution (this is of course equivalent to solving in y the equation $x = y + tu_0(y)$ as above).

3.2   Eikonal Equations

We consider in $\mathbf{R}_x^n$ fully nonlinear equations of the form

$$F(x, \nabla u(x)) = 0,$$

where $F : \mathbf{R}_x^n \times \mathbf{R}_\xi^n \supset \Omega \to \mathbf{R}$ is a given C^∞ function. We look for a C^1 real solution $u : \mathbf{R}^n \supset \omega \to \mathbf{R}$.

Definition 3.5. *The Hamiltonian field H_F associated to F is the field on $\Omega \subset \mathbf{R}_x^n \times \mathbf{R}_\xi^n$ defined by*

$$H_F(x, \xi) = \Sigma(\partial_{\xi_i} F)\partial_{x_i} - \Sigma(\partial_{x_i} F)\partial_{\xi_i}.$$

Note that, by Theorem 1.10 of chapter 1, F is constant along the integral curves of H_F, since $H_F F = 0$. If the equation is linear, that is, if $F(x, \xi) = \Sigma a_i(x)\xi_i$, the Hamiltonian field is simply

$$H_F = \Sigma a_i(x)\partial_{x_i} - \Sigma(\partial_{x_i} F)\partial_{\xi_i},$$

projecting as a onto $\mathbf{R}_x^n$.

For simplicity, we first explain the method of characteristics for an equation in the plane $\mathbf{R}^2_{x,y}$

$$F(x, y, \partial_x u, \partial_y u) = 0.$$

We assume here that $\partial_\xi F, \partial_\eta F$ do not both vanish. Just as in the quasi-linear case, we reduce our problem to a purely *geometric* problem. To a couple of C^1 real functions p, q on $\omega \subset \mathbf{R}^2_{x,y}$ we associate the 2-manifold in $\omega \times \mathbf{R}^2_{\xi,\eta} \subset \mathbf{R}^4$ defined by

$$S = \{(x, y, \xi, \eta), (x, y) \in \omega, \ \xi = p(x, y), \ \eta = q(x, y)\}.$$

The method is based on the following observation:

Observation 3.6. *Suppose $F = 0$ on S. Then $\partial_y p = \partial_x q$ in ω if and only if the Hamiltonian field H_F is tangent to S.*

In fact, H_F tangent to S means, on S, $H_F(\xi - p) = 0, H_F(\eta - q) = 0$, that is

$$\partial_x F + \partial_\xi F \partial_x p + \partial_\eta F \partial_y p = 0, \ \ \partial_y F + \partial_\xi F \partial_x q + \partial_\eta F \partial_y q = 0.$$

On the other hand, differentiating $F(x, y, p(x, y), q(x, y)) = 0$ with respect to x and y yields

$$\partial_x F + \partial_\xi F \partial_x p + \partial_\eta F \partial_x q = 0, \ \ \partial_y F + \partial_\xi F \partial_y p + \partial_\eta F \partial_y q = 0.$$

Subtracting the above two equations, we obtain

$$\partial_\xi F(\partial_y p - \partial_x q) = 0, \ \ \partial_\eta F(\partial_y p - \partial_x q) = 0,$$

which yields the observation.   □

Consider now the Cauchy problem with data u_0 given on the initial hypersurface $\Sigma_0 = \{(x, y), y = 0\}$. Suppose there exists a solution $u \in C^2(\omega)$ of the Cauchy problem

$$F(x, y, \partial_x u, \partial_y u) = 0, u(x, 0) = u_0(x).$$

Then, for $y = 0$,
$$F(x, 0, u_0'(x), \partial_y u(x, 0)) = 0.$$

To handle the Cauchy problem with data u_0 on Σ_0, we give a real $u_0 \in C^2$ on Σ_0 and make the following assumption:

Assumption 3.7. *There exists a C^1 real function $\eta(x)$ with*

$$F(x, 0, u_0'(x), \eta(x)) = 0.$$

We see that $\eta(x)$ is a "candidate" for the future $\partial_y u(x,0)$. Defining then the 1-manifold $\Sigma \subset \mathbf{R}^4$ by

$$\Sigma = \{(x,0,u_0'(x),\eta(x))\},$$

we have to solve the following **geometric problem**: Find a 2-submanifold $S \subset \mathbf{R}^4$ such that the following conditions are met:

i) S is the graph of a C^1 map $\Omega \ni (x,y) \mapsto (\xi = p(x,y), \eta = q(x,y))$;

ii) S contains Σ;

iii) $F = 0$ on S;

iv) H_F is tangent to S.

If we can solve this problem, we know that $\partial_y p = \partial_x q$ in Ω. If Ω is a disc, for instance (or more generally a star-shaped domain), there will be $v \in C^2(\Omega)$ with

$$\partial_x v = p, \ \partial_y v = q.$$

For $y = 0$, we have then

$$p(x,0) = (\partial_x v)(x,0) = \partial_x(v(x,0)) = u_0'(x), \ (\partial_y v)(x,0) = \eta(x).$$

Adjusting v by a constant, we obtain our solution u with

$$u(x,0) = u_0(x), \ (\partial_y u)(x,0) = \eta(x).$$

Just as before, condition (iv) implies that integral curves of H_F starting from Σ stay on S. Hence, we take for S the union of all integral curves of H_F starting from Σ. However, we have to be careful about the following questions:

1. Is S a submanifold?

2. Is S the graph of a C^1 couple (p,q)?

3. Does F vanish on S?

Again, the answers to questions 1 and 2 are in general only local:

1. and 2. Let $m_0 = (x_0, 0, u_0'(x_0), \eta(x_0)) \in \Sigma$, and assume $\partial_\eta F(m_0) \neq 0$. Then H_F is not tangent to Σ, hence S is a manifold (see Appendix, A.2); moreover, by the same argument as in Section 3.1, $T_{m_0}S$ cannot contain a

vertical vector, hence S is the graph of a C^1 map (Again, see Appendix, Theorem A.14).

3. By construction, $F = 0$ on Σ, and $H_F F = 0$. Hence, condition (iii) is satisfied by our choice of S.

We have proved the following theorem:

Theorem 3.8. *Let $x_0 \in \mathbf{R}$ and assume that there exists $\eta_0 \in \mathbf{R}$ such that, at $m_0 = (x_0, 0, u_0'(x_0), \eta_0)$,*

$$F(m_0) = 0, \partial_\eta F(m_0) \neq 0.$$

Then there is a C^1 real function η near x_0 in $\mathbf{R}_x$ such that

$$F(x, 0, u_0'(x), \eta(x)) = 0, \ \eta(x_0) = \eta_0.$$

There is a unique solution $u \in C^2$ in a neighborhood of $(x_0, 0)$ of the Cauchy problem

$$F(x, y, \partial_x u(x,y), \partial_y u(x,y)) = 0, \ u(x,0) = u_0(x), \ (\partial_y u)(x,0) = \eta(x).$$

The graph of ∇u is the union of the integral curves of H_F starting from

$$\Sigma = \{(x, 0, u_0'(x), \eta(x))\}.$$

To complete the proof, it remains for us to observe that the existence of the function η is an immediate consequence of the implicit function theorem. An equivalent approach is as follows: Use the implicit function theorem to rewrite the equation $F(x, y, \partial_x u, \partial_y u) = 0$, for $(x, y, \partial_x u, \partial_y u)$ in a neighborhood of $(x_0, 0, u_0'(x_0), \eta_0)$ as $\partial_y u = G(x, y, \partial_x u)$ for some C^∞ function G. Theorem 3.8 shows that the Cauchy problem has as many branches of solutions as the number of roots η_0 we can find.

We turn now to the general case of an equation $F(x, \nabla u(x)) = 0$ for $x \in \mathbf{R}^n$, where the construction is exactly the same. We write $x = (x_1, \ldots, x_{n-1}, x_n) = (x', x_n) \in \mathbf{R}_{x'}^{n-1} \times \mathbf{R}_{x_n}$, and we want to solve the Cauchy problem with data u_0 on $\Sigma_0 = \{x_n = 0\}$, close to $(x_0', 0)$. We assume that there is a real η_0 such that, at $m_0 = (x_0', 0, \nabla_{x'} u_0(x_0'), \eta_0)$,

$$F(m_0) = 0, \ \partial_{\xi_n} F(m_0) \neq 0. \ .$$

We thus obtain locally near x_0' a C^1 real function $\eta(x')$ satisfying

$$F(x', 0, \nabla_{x'} u_0(x'), \eta(x')) = 0, \ \eta(x_0') = \eta_0.$$

Defining the $(n-1)$-submanifold $\Sigma \subset \mathbf{R}_x^n \times \mathbf{R}^n$ by

$$\Sigma = \{(x', 0, \nabla_{x'} u_0(x'), \eta(x'))\},$$

we take for S the union of all integral curves of H_F starting from Σ. Then, locally near m_0, $S = \{\xi = p(x)\}$ for some

$$p : \mathbf{R}_x^n \supset \Omega \to \mathbf{R}_\xi^n$$

of class C^1.

Claim: The functions $\omega_{ij} = \partial_j p_i - \partial_i p_j$ all vanish.

Admitting this claim, we obtain, if Ω is a ball (or more generally a star-shaped domain), $p = \nabla u$ for some $u \in C^2$, with the desired properties. To prove the claim, we establish that the functions ω_{ij} satisfy a linear homogeneous system of ordinary differential equations (ODE) and all vanish on Σ_0:

- Let us write that H_F is tangent to S: For all i, $H_F(\xi_i - p_i(x)) = 0$, which gives

$$\partial_i F(x, p(x)) + \Sigma(\partial_{\xi_j} F)(x, p(x))\partial_j p_i(x) = 0, \ i = 1, \dots, n.$$

Differentiating this with respect to x_k, we obtain (all derivatives of F being taken at the point $(x, p(x))$)

$$\partial_{ik}^2 F + \Sigma(\partial_{i\xi_j}^2 F)\partial_k p_j + \Sigma(\partial_{k\xi_j}^2 F)\partial_j p_i$$
$$+ \Sigma(\partial_{\xi_j \xi_l}^2 F)(\partial_k p_l)(\partial_j p_i) + \Sigma(\partial_{\xi_j} F)\partial_{jk}^2 p_i = 0.$$

Subtracting the similar equality with i and k exchanged, we get with $X = \Sigma(\partial_{\xi_j} F)(x, p(x))\partial_j$,

$$X\omega_{ik} + \Sigma(\partial_{i\xi_j}^2 F)\omega_{jk} + \Sigma(\partial_{k\xi_j}^2 F)\omega_{ij} + \Sigma(\partial_{\xi_j \xi_l}^2 F)(\omega_{lk}\partial_j p_i - \omega_{li}\partial_j p_k) = 0.$$

This is our system of ODE.

- By definition of the submanifold Σ,

$$p(x', 0) = (\partial_1 u_0(x'), \dots, \partial_{n-1} u_0(x'), \eta(x')).$$

Differentiating the equation satisfied by η with respect to x_i $(i < n)$,

$$\partial_i F + \Sigma_{j<n}(\partial_{\xi_j} F)\partial_{ij}^2 u_0 + (\partial_{\xi_n} F)\partial_i \eta = 0.$$

Comparing with the above equation $H_F(\xi_i - p_i) = 0$ for $x_n = 0$, we obtain $\partial_i p_n(x', 0) = \partial_n p_i(x', 0)$. Hence all ω_{ij} vanish on Σ_0, and satisfy the linear homogeneous system of ODE above. The claim is proved. $\qquad\square$

Example 3.9. Let Σ_0 be a smooth curve in the plane and consider the Cauchy problem

$$F(x, y, \partial_x u, \partial_y u) = (\partial_x u)^2 + (\partial_y u)^2 - 1 = 0,$$

with $u = 0$ on Σ_0. Then

$$H_F = 2\xi \partial_x + 2\eta \partial_y.$$

The manifold Σ above Σ_0 is the collection of unit normals to Σ_0, and an integral curve of H_F starting from a point of Σ projects on the normal to Σ_0, the gradient of u being constant along this curve and equal to the unit normal. Hence u is just the distance to Σ_0 locally. Too far away from Σ_0, the distance function can become nonsmooth: for instance, the distance to a circle ceases being C^1 at the center.

Example 3.10. Let

$$F = \xi_n^2 - (\xi_1^2 + \cdots + \xi_{n-1}^2) = 0.$$

To solve the eikonal equation with data $u(x', 0) = \omega \cdot x'$ (ω being a parameter), we have two choices:

$$\partial_n u = \pm |\nabla_{x'} u|, \text{ or } \partial_n u(x', 0) = \pm |\omega|.$$

The solutions are easily seen to be $u(x) = x' \cdot \omega \pm x_n |\omega|$.

3.3 Exercises

1. Compute explicitly the solutions of Burgers equation close to the origin for the data
$$u_0(x) = x^2 + bx + c, \ u_0(x) = (1 + x)^{-1}.$$

2.(a) Let $u_0 \in C^1(\mathbf{R})$ be a real function vanishing outside $[a, b]$. Prove that for all
$$0 \leq t < \bar{T} \equiv (\max -u_0')^{-1},$$
the map $y \mapsto y + t u_0(y)$ is a strictly increasing bijection of $\mathbf{R}$ onto itself. Deduce from this that there exists a unique C^1 solution u of the Cauchy problem for Burgers equation

$$\partial_t u + u \partial_x u = 0, \ u(x, 0) = u_0(x), \ 0 \leq t < \bar{T}.$$

(b) Suppose now $u_0' \geq 0$ everywhere. Show that Burgers equation has then a global unique solution for $t \geq 0$.

3.(a) Consider

$$u_0(x) = -ax + b\frac{x^3}{3}, \ a > 0, \ b > 0,$$

and set $\bar{T} = 1/a$. Display, according to the three cases $0 \le t < \bar{T}, t = \bar{T}$, $t > \bar{T}$, the behavior of the function $f_t(y) = y + tu_0(y)$. Show that there exists for $t \ge \bar{T}$ an increasing function $c(t) \ge 0$, with $c(\bar{T}) = 0, c'(\bar{T}) = 0$, such that inside the cusp region

$$\{(x,t), t \ge \bar{T}, -c(t) \le x \le c(t)\},$$

the equation (in y) $y + tu_0(y) = x$ has three solutions:

$$y_1(x,t) \le y_2(x,t) \le y_3(x,t).$$

Note that $y_3(x,t)$ is in fact defined in the region

$$\{t < \bar{T}\} \cup \{(x,t), t \ge \bar{T}, x > -c(t)\},$$

while $y_1(x,t)$ is defined in the region

$$\{t < \bar{T}\} \cup \{(x,t), t \ge \bar{T}, x < c(t)\}.$$

(b) Consider the Cauchy problem for Burgers equation with the initial data

$$u_0(x) = -ax + b\frac{x^3}{3}, \ a > 0, \ b > 0$$

as in 3.(a), and let S be the surface $\{z = u(x,t)\}$ constructed by the method of characteristics. Display, according to the values of t, the behavior, for x close to zero, of the curve

$$\{(x,z), (x,t,z) \in S\}.$$

Show that the projection onto the (x,t) plane of the subset of points m of S where $T_m S$ contains vertical vectors is just the cusp $\{(x,t), t \ge \bar{T}, x = \pm c(t)\}$ from 3.(a).

4. Let $u_0(x)$ be 1 for $x \le -1$, -1 for $x \ge 1$, and linear in between. Compute explicitly the solution u of the Cauchy problem for Burgers equation

$$\partial_t u + u\partial_x u = 0, \ u(x,0) = u_0(x), \ 0 \le t \le 1.$$

What happens for $t = \bar{T} = 1$? Why is this in constrast with the "generic" case displayed in Exercise 3.(a)?

5.(a) Let $u \in C^2(\mathbf{R} \times [0, T[)$ be a real solution of the Cauchy problem for Burgers equation

$$\partial_t u + u \partial_x u = 0, \ u(x, 0) = u_0(x).$$

Let $(x(s) = x_0 + s u_0(x_0), s)$ be the integral curve of $\partial_t + u \partial_x$ starting from $(x_0, 0)$. Show that the function $q(s) = (\partial_x u)(x(s), s)$ satifies $q' + q^2 = 0$. Compute q explicitly. If $u_0'(x_0) < 0$, q becomes infinite at $s = s(x_0) > 0$. Deduce from this that

$$T \leq \bar{T} \equiv (\max - u_0')^{-1}.$$

Taking Exercise 2 into account, we see that for a given $u_0 \in C^2$ there exists a C^2 solution u of the Cauchy problem exactly in the strip $\{0 \leq t < \bar{T}\}$. The number $\bar{T}$ is called the **lifespan** of the smooth solution. In the sequence we assume $T = \bar{T}$.

(b) Show that the points x of $\{t = \bar{T}\}$ where $\partial_x u$ blows up are exactly the critical values of the function

$$y \mapsto y + \bar{T} u_0(y).$$

Deduce from Sard's theorem that they form a set of measure zero.

(c) Let x_m be a minimum of u_0' with $u_0'(x_m) < 0$. Let $(x(t), t)$ be the characteristic starting from $(x_m, 0)$, and $M_0 = (x(s(x_m)), s(x_m))$. Prove that in a neighborhood of M_0 in $\{t \leq \bar{T}\}$, $\partial_x u$ is negative and $\max |(\partial_x u)(., t)| = (\bar{T} - t)^{-1}$.

(d) Using for $t < \bar{T}$ the formula $u(y + t u_0(y), t) = u_0(y)$ for the solution u, prove that

$$\int |\partial_x u(x, t)| dx = \int |u_0'(x)| dx.$$

6.(a) Let

$$a : \mathbf{R_u} \to \mathbf{R}^n, \ a(u) = (a_1(u), \ldots, a_n(u))$$

be a C^2 function, and fix $u_0 \in C^1(\mathbf{R}_x^n)$, a bounded real function. Define

$$\Psi_t : \mathbf{R}^n \to \mathbf{R}^n, \ \Psi_t(y) = y + t a(u_0(y)).$$

Compute the differential $D_{x_0} \Psi_t$. Define $\bar{T}$ by

$$\bar{T}^{-1} = \max - \Sigma a_i'(u_0(y))(\partial_i u_0)(y).$$

Show that if $t < \bar{T}$ this differential is invertible. Using Hadamard's Theorem, prove that Ψ_t is a global diffeomorphism from $\mathbf{R}^n$ onto itself.

(b) Let $u \in C^2(\mathbf{R}_x^n \times [0, T[)$ be a real solution of the quasilinear Cauchy problem

$$\partial_t u + \Sigma a_i(u)\partial_i u = 0, \ u(x, 0) = u_0(x).$$

Show that the integral curve of $\partial_t + \Sigma a_i(u)\partial_i$ starting from $(x_0, 0)$ is

$$t \mapsto (x(t), t), \ x(t) = x_0 + ta(u_0(x_0)).$$

Set $d = \Sigma \partial_i(a_i(u))$. Prove that

$$(\partial_t + \Sigma a_i(u)\partial_i)d + d^2 = 0.$$

Deduce from this that $T \leq \bar{T}$.

(c) Assume now that the maximum defining $(\bar{T})^{-1}$ is attained at some point x_0. Deduce from (a) and (b) that there exists a solution u of the Cauchy problem in the strip $0 \leq t < \bar{T}$, with $u(\Psi_t(y), t) = u_0(y)$, and that d becomes infinite at the point $m_0 = \Psi_{\bar{T}}(x_0)$.

(d) Let $n = 2$ for simplicity. Show that there exists a nonzero vector field X with no t component such that Xu is bounded near m_0 (in other words, not all components of ∇u blow up).

7.(a) Let $u \in C^2$ be a real solution of the Cauchy problem for Burgers equation

$$\partial_t u + u\partial_x u = 0, \ u(x, 0) = u_0(x).$$

The time at which $\partial_x u$ blows up along the characteristic starting from x_0 is $s(x_0) = -1/u_0'(x_0)$ (see Exercise 5.(a)). Assuming $u_0'(x_0) < 0, u_0''(x_0) \neq 0$, describe the set

$$\gamma = \{(x, t), x = y + s(y)u_0(y), t = s(y)\}$$

for y close to x_0. Show that γ is a smooth curve, and an envelope of the characteristics.

(b) Assume now

$$u_0'(x_0) < 0, \ u_0''(x_0) = 0, \ u_0'''(x_0) > 0.$$

Describe the set γ and show that it is the envelope of the characteristics. Prove that γ is the boundary of the cusp region discussed in Exercise 3.

8. Let $u \in C^3$ be a real solution of Burgers equation for $0 \leq t < \bar{T} \equiv (\max(-u_0'))^{-1}$. Compute explicitly the function

$$r(t) = (\partial_x^2 u)(x_0 + tu_0(x_0), t), \ t < \bar{T}.$$

What is the blowup rate of $\partial_x^2 u$ when $t \to \bar{T}$?

9. Consider the scalar equation

$$\partial_t u + a(u)\partial_x u = 0,$$

where $a \in C^\infty(\mathbf{R})$ and $u \in C^1$ are real. Prove that one can reduce the above to Burgers equation by setting $v = a(u)$. In particular, compute the blowup time $\bar{T}$ for a given real initial data $u(x, 0) = u_0(x)$, $u_0 \in C_0^1$.

10. Let $u \in C^1(\mathbf{R}_x \times [0, T[)$ be a real solution of the equation

$$\partial_t u + a(x, t, u)\partial_x u = f(u),$$

where a and f are C^∞ and real, $f(0) = 0$. Assume that $u_0(x) = u(x, 0)$ vanishes outside $[a, b]$. What can be said about the support of u?

11. Consider the Cauchy problem in the plane

$$\partial_t u + u\partial_x u = u^2, \ u(x, 0) = u_0(x),$$

where $u_0 \in C_0^2(\mathbf{R})$ is real and not identically zero.

(a) Show the inequalities $\max(u_0 - u_0') > 0$, $\max(u_0 - u_0') \geq \max u_0$. Prove that if there is x_0 where

$$u_0(x_0) = \max u_0 > 0, \ u_0''(x_0) < 0,$$

then $\max(u_0 - u_0') > \max u_0$.

(b) Use the method of characteristics to solve the equation. Prove that a C^2 solution u exists for

$$0 \leq t < \bar{T} \equiv (\max(u_0 - u_0'))^{-1}.$$

(c) Let $u \in C^2(\mathbf{R} \times [0, T[)$ be a solution of the Cauchy problem. Denote by $(x(t), t)$ the characteristic starting from $(x_0, 0)$ and $q(t) = (\partial_x u)(x(t), t)$. Establish the ODE satisfied by q, and compute q explicitly. Deduce from this that $T \leq \bar{T}$ (hence, as in Exercise 5, $\bar{T}$ is the lifespan).

12.(a) Let $F(x, \xi)$ $(x \in \mathbf{R}^n, \ \xi \in \mathbf{R}^n)$ be C^∞, real, and (positively) homogeneous with respect to ξ. Show that a C^1 solution u of the eikonal equation $F(x, \nabla u) = 0$ is constant along the characteristics.

(b) Let $S \subset \mathbf{R}^n$ be a $(n-1)$-submanifold defined by an equation $\{f = 0\}$ (with $\nabla f \neq 0$). Assume F as in (a) and also

$$x \in S \Rightarrow F(x, \nabla f(x)) = 0.$$

(If F happens to be the symbol of a differential operator P, we say that the surface S is characteristic for P.) Show that an integral curve $(x(s), \xi(s))$ of H_F with $x(0) \in S$, $\xi(0) = \nabla f(x(0))$, satisfies

$$x(s) \in S, \ \xi(s) = \nabla f(x(s)).$$

13.(a) Consider the Cauchy problem in the plane

$$G(x, t, \partial_x \phi, \partial_t \phi) \equiv \partial_t \phi - F(x, t, \partial_x \phi) = 0, \ \phi(x, 0) = \phi_0(x).$$

Compute H_G and explain how the graph $\tilde{S} \subset \mathbf{R}^4$ of $\nabla \phi$ is constructed by the method of characteristics.

(b) Set $\psi = \partial_x \phi$. What is the new Cauchy problem for ψ corresponding to that for ϕ? Suppose the Cauchy problem for ψ is solved by the method of characteristics using a field V (see Section 3.1) and let $S \in \mathbf{R}^3$ be the graph of ψ. Explain how the solution ϕ can be recovered from ψ. Show that S is the projection of $\tilde{S}$ and V the projection of H_G.

3.4   Notes

The material in this chapter can be found in many places, for instance Courant-Hilbert [8], Evans [9], John [12], Taylor [23], and Vainberg [24]. We concentrated on smooth solutions of the Cauchy problem, with no attempt to discuss nonsmooth solutions or other boundary value problems for Hamiton–Jacobi equations. Some hints about blowup of smooth solutions are given in the Exercises.

Chapter 4

Conservation Laws in One-Space Dimension

4.1 First Definitions and Examples

Definition 4.1. *Conservation laws in the plane* $\mathbf{R}^2_{x,t}$ *are special nonlinear systems in divergence form*

$$\partial_t u + \partial_x(F(u)) = 0.$$

Here, $(x,t) \in \mathbf{R}^2$ *are the coordinates in the plane,* $u : \mathbf{R}^2 \supset \Omega \to \mathbf{R}^N$ *is an unknown vector function, and* $F : \mathbf{R}^N \to \mathbf{R}^N$ *is a* C^∞ *given function.*

If $u \in C^1$, the system can be equivalently written as

$$\partial_t u + F'(u)\partial_x u = 0,$$

and this is a **quasilinear system** on u. In accordance with the assumptions we made in Chapters 2 and 3, we will always assume that the $N \times N$ matrix $A(u) = F'(u)$ has real and distinct eigenvalues

$$\lambda_1(u) < \cdots < \lambda_N(u),$$

with right and left eigenvectors $r_j(u)$, $\ell_j(u)$:

$$A(u)r_j(u) = \lambda_j(u)r_j(u), \ {}^t\ell_j(u)A(u) = \lambda_j(u){}^t\ell_j(u).$$

We say that we have a **strictly hyperbolic system** of conservation laws. In the sequence, all systems will be assumed to be strictly hyperbolic.

S. Alinhac, *Hyperbolic Partial Differential Equations*, Universitext,
DOI 10.1007/978-0-387-87823-2_4, © Springer Science+Business Media, LLC 2009

The simplest example is Burgers equation, written as

$$\partial_t u + \partial_x \left(\frac{u^2}{2} \right) = 0.$$

The next example in simplicity is the so-called **p-system**: consider a nonlinear second order equation in the plane of the form

$$\partial_t^2 \phi - \partial_x [p(\partial_x \phi)] = 0,$$

where $p : \mathbf{R} \to \mathbf{R}$ is given and ϕ is real. Transforming it into a first order system, we set

$$u_1 = \partial_x \phi, u_2 = \partial_t \phi.$$

We thus obtain the p-system

$$\partial_t u_1 - \partial_x u_2 = 0, \ \partial_t u_2 - \partial_x [p(u_1)] = 0.$$

This system is stricly hyperbolic if $p' > 0$.

A physical example is given by the complete **Euler system**:

$$\partial_t \rho + \partial_x (\rho u) = 0, \ \ \partial_t (\rho u) + \partial_x (\rho u^2 + p) = 0,$$

$$\partial_t [\rho(u^2/2 + e)] + \partial_x [\rho u (u^2/2 + e + p/\rho)] = 0.$$

Here $\rho > 0$ is the density of the fluid, $u \in \mathbf{R}$ is its velocity, p its pressure, and e its internal energy. The function $e = e(\rho, p)$ is known from physical considerations about the nature of the fluid, thus (ρ, u, p) are the unknowns of this 3×3 system.

We remark that the systems that we consider here, being written in *divergence form*, that is, with the derivatives *before* the nonlinear terms, admit interesting (continuous or not) solutions that are not C^1. We first display examples of such solutions.

4.2   Examples of Singular Solutions

4.2.1. Shocks
Let $\gamma = \{(x,t), x = \phi(t)\}$ be a C^1 curve, and u_r and u_l be C^1 functions respectively for $x \geq \phi(t)$ and $x \leq \phi(t)$. The function u defined (almost everywhere) to be u_r for $x > \phi(t)$ and u_l for $x < \phi(t)$, discontinuous across the shock curve γ, is called a *shock*. The following theorem characterizes the

shocks that are solutions, in the sense of distribution theory, of a given system of conservation laws. For people not familiar with distribution theory, the theorem can be admitted and used as a definition of a shock solution.

Theorem 4.2 (Rankine–Hugoniot relation). *A shock u is a solution (in the sense of distribution theory) of the system of conservation laws*

$$\partial_t u + \partial_x(F(u)) = 0$$

if and only if

i) u_r and u_l are solutions of the system on either side;

ii) the Rankine–Hugoniot relation

$$[F(u_r) - F(u_l)](\phi(t), t) = \phi'(t)[u_r - u_l](\phi(t), t)$$

holds on γ (u_r and u_l being the corresponding right and left limits).

Proof: To prove the theorem without relying too much on distribution theory, we approximate u and $F(u)$ as follows:

$$u_\epsilon = \tilde{u}_r H_\epsilon(x - \phi(t)) + \tilde{u}_l H_\epsilon(-(x - \phi(t))),$$
$$F_\epsilon = F(\tilde{u}_r) H_\epsilon(x - \phi(t)) + F(\tilde{u}_l) H_\epsilon(-(x - \phi(t))).$$

Here, $\tilde{u}_r$ and $\tilde{u}_l$ are C^1 extensions of u_r and u_l beyond γ; $H_\epsilon(s)$ is an approximation of the Heaviside function H, defined by

$$H_\epsilon(s) = \frac{1}{2}\left(1 + K\left(\frac{s}{\epsilon}\right)\right),$$

K being an odd C^∞ increasing function with limits ± 1 at $s = \pm\infty$. Thus H'_ϵ is an even function converging to the Dirac mass δ as $\epsilon \to 0$. In the sense of distributions,

$$u_\epsilon \to u, \quad F_\epsilon \to F(u),$$

hence $\partial_t u_\epsilon \to \partial_t u$, $\partial_x F_\epsilon \to \partial_x F(u)$. Now, in the classical sense,

$$\partial_t u_\epsilon = (\partial_t \tilde{u}_r) H_\epsilon(x - \phi(t)) + (\partial_t \tilde{u}_l) H_\epsilon(-(x - \phi(t))) - \phi'(t) H'_\epsilon(\tilde{u}_r - \tilde{u}_l),$$
$$\partial_x F_\epsilon = (\partial_x F(\tilde{u}_r)) H_\epsilon(x - \phi(t)) + (\partial_x F(\tilde{u}_l)) H_\epsilon(-(x - \phi(t)))$$
$$+ H'_\epsilon(F(\tilde{u}_r) - F(\tilde{u}_l)).$$

Hence, in the sense of distributions,

$$\partial_t u_\epsilon + \partial_x F_\epsilon \to \partial_t u + \partial_x(F(u)) = \delta(x - \phi(t))[F(u_r) - F(u_l) - \phi'(t)(u_r - u_l)],$$

which proves the theorem. $\qquad\square$

4.2.2. Examples for Burgers Equation

For Burgers equation

$$\partial_t u + \partial_x \left(\frac{u^2}{2}\right) = 0,$$

the Rankine–Hugoniot relation simply reads $\phi'(t) = \frac{1}{2}(u_r + u_l)(\phi(t), t)$.

Example 4.3. Take $\phi(t) = t/2$, $u_r \equiv 0$ and $u_l \equiv 1$: this is a piecewise constant shock solution. The shock front $x = t/2$ moves to the right with speed $1/2$; the speed u_l to the left is greater than the shock speed, which is in turn greater than the speed u_r to the right. It is like a breaking wave. Such a shock is called a *compressive shock*.

Example 4.4. Take now $\phi(t) = t/2$, $u_r \equiv 1$, and $u_l \equiv 0$; this is again a piecewise constant shock solution. But now, the speed to the right is greater than the shock speed: the particles move away from the shock.

We will see below in this chapter that compressive shocks are physically (and mathematically) admissible, while "rarefaction shocks" as in example 4.4 are not admissible.

4.2.3. Rarefaction waves

We give only one example for Burgers equation and will come back to the subject in Section 4.4 for general systems. Consider the function $u(x, t)$ which is 0 for $x \leq 0$, 1 for $x \geq t$, and x/t in between. This is a continuous function for $t > 0$, and

$$\partial_t \left(\frac{x}{t}\right) + \left(\frac{x}{t}\right) \partial_x \left(\frac{x}{t}\right) = -\frac{x}{t^2} + \frac{x}{t^2} = 0,$$

which proves that u is a solution of Burgers equation, called a *(centered) rarefaction wave*.

We remark that the Cauchy data for this solution u (the Heaviside function) are the same as for the shock solution of Example 4.4 above: We thus have two solutions of Burgers equation with the same Cauchy data; this is an embarrassing mistake. We will see later for what reasons we will discard the bad shock of Example 4.4, and keep the rarefaction wave u as the admissible solution for these data.

4.3   Simple Waves

In this section and Sections 4.4 and 4.5 only, we enlarge the discussion to general quasilinear systems of the form

$$\partial_t u + A(u)\partial_x u = 0,$$

where $A(u)$ is an $N \times N$ matrix depending smoothly on u. We assume as before that A has real and distinct eigenvalues

$$\lambda_1(u) < \cdots < \lambda_N(u),$$

with corresponding right and left eigenvectors $r_j(u)$, $\ell_j(u)$ $(j = 1, \ldots, N)$.

Definition 4.5. *A C^1 simple wave in $\Omega \subset \mathbf{R}^2_{x,t}$ is a solution u of the system of the form*

$$u(x,t) = U(\psi(x,t)),$$

where $U : \mathbf{R} \supset I \to \mathbf{R}^N$ is a C^1 curve defined on some real interval and $\psi : \Omega \to I$ is a C^1 function.

In other words, a simple wave solution has its values on a curve (the image of U). It can be thought of as an intermediate case between constant solutions (values at one point) and general solutions (values on a 2-surface of $\mathbf{R}^N$)). Since both U and ψ are C^1,

$$\partial_t u = U'(\psi)\partial_t\psi, \ \partial_x u = U'(\psi)\partial_x\psi,$$
$$\partial_t u + A(u)\partial_x u = [\partial_t\psi + A(u)\partial_x\psi]U'(\psi).$$

For u to be a solution of the system, it is enough (and almost necessary) to take, for some j,

$$U'(s) = r_j(U(s)), \ \partial_t\psi + \lambda_j(U(\psi))\partial_x\psi = 0.$$

We take these relations as the definition of a j-**simple wave**.

What we have gained is this: We obtain a solution of the system by computing an integral curve of r_j (that is, solving a system of ODE) and solving a *scalar equation* for ψ. From this construction, we obtain N families of simple waves, one for each mode λ_j. Note that for a scalar equation, all waves are simple.

Example 4.6. Consider a 2-system in diagonal form

$$\partial_t u_1 + \lambda_1(u)\partial_x u_1 = 0, \ \partial_t u_2 + \lambda_2(u)\partial_x u_2 = 0.$$

Here, the matrix A is diagonal, the r_j form the standard basis of $\mathbf{R}^2$. A 1-simple wave means that u_2 is a constant, a 2-simple wave that u_1 is a constant.

4.4   Rarefaction Waves

The structure of the rarefaction solution of Burgers equation given in Section 4.2 may seem very special, but strikingly enough, we will show how to construct such solutions for general systems!

Definition 4.7. *A (centered) rarefaction wave is a solution of the system*

$$\partial_t u + A(u)\partial_x u = 0$$

with the following structure:

i) *For some $s_l < s_r$, the solution u is constant with value $u_l \in \mathbf{R}^N$ for $x \le s_l t$, constant with value $u_r \in \mathbf{R}^N$ for $x \ge s_r t$,*

ii) *The exists $U \in C^1([s_l, s_r], \mathbf{R}^N)$ with $U(s_l) = u_l, U(s_r) = u_r$, such that for $s_l t \le x \le s_r t$,*

$$u(x, t) = U(x/t).$$

From the definition, we see that a rarefaction wave is a special case of simple wave (though not quite C^1), with $\psi = x/t$. For u to be a j-simple wave, the function $\psi(x, t) = x/t$ has to satisfy

$$\partial_t \psi + \lambda_j(U(\psi))\partial_x \psi = 0;$$

that is, $\lambda_j(U(s)) = s$, and in particular, $\lambda_j(u_l) = s_l, \lambda_j(u_r) = s_r$. This leads us naturally to a new definition.

Definition 4.8. *The eigenvalue $\lambda_j(u)$ is genuinely nonlinear if*

$$r_j(u) \cdot \nabla \lambda_j(u) \ne 0.$$

In this case, r_j is said to be normalized if it is chosen so that

$$r_j(u) \cdot \nabla \lambda_j(u) \equiv 1.$$

The eigenvalue λ_j is linearly degenerate if

$$r_j(u) \cdot \nabla \lambda_j(u) \equiv 0.$$

For instance, for the Burgers equation, $r_1(u) = 1, r_1 \nabla \lambda_1 = 1$. In the case of the 2-system in diagonal form considered in Section 4.3, λ_1 is genuinely nonlinear if and only if $\partial_1 \lambda_1 \ne 0$.

The following theorem gives the existence of rarefaction waves.

Theorem 4.9. *Assume that, for some j, λ_j is genuinely nonlinear, and let r_j be the corresponding normalized eigenvector. Let u_l be a given constant state in $\mathbf{R}^N$, and define $s_l = \lambda_j(u_l)$, and U by*

$$U'(s) = r_j(U(s)), \; U(s_l) = u_l.$$

For some $s_r > s_l$, set $u_r = U(s_r)$. Then the function u, which is u_l for $x \le s_l t$, u_r for $x \ge s_r t$, and $U(x/t)$ in between, is a rarefaction wave solution of the given system.

To prove the theorem, it is enough to note that $\lambda_j(U(s)) \equiv s$, since there is equality for $s = s_l$ and

$$\frac{d}{ds}[\lambda_j(U(s))] = U'(s) \cdot \nabla\lambda_j(U(s)) = (r_j \cdot \nabla\lambda_j)(U(s)) \equiv 1. \qquad \square$$

We say that we have connected the constant state u_r to u_l by a j-rarefaction wave. Note that this is possible only if u_r belongs to the *half*-integral curve of r_j through u_l indicated by the direction of r_j. We call this half-curve the j-**rarefaction curve from** u_l.

4.5   Riemann Invariants

Consider a quasilinear strictly hyperbolic system $\partial_t u + A(u)\partial_x u = 0$ as in Section 4.3.

Definition 4.10. *A j-Riemann invariant is a C^1 real function R on $\Omega \subset \mathbf{R}^N_u$ such that*

$$r_j(u) \cdot \nabla R(u) \equiv 0.$$

In other words, the function R is constant along the integral curves of r_j. One striking application of this concept is the diagonalization of 2×2 systems.

Theorem 4.11. *Consider a quasilinear 2×2 system, and let $R_1(u)$ and $R_2(u)$ be 1- and 2-Riemann invariants defined on $\Omega \subset \mathbf{R}^2$. Assume that*

$$K : (u_1, u_2) \mapsto (v_1 = R_1(u), v_2 = R_2(u))$$

is a diffeomorphism from Ω onto Ω'. Then, for a C^1 function u with values in Ω, the system $\partial_t u + A(u)\partial_x u = 0$ is equivalent to the diagonal system

$$\partial_t v_1 + \bar{\lambda}_2(v)\partial_x v_1 = 0, \; \partial_t v_2 + \bar{\lambda}_1(v)\partial_x v_2 = 0,$$

where

$$\bar{\lambda}_j(v) = \lambda_j(K^{-1}v).$$

To prove this, remember that ${}^t\ell_j r_k = 0$ when $j \neq k$. Hence ∇R_2, which is orthogonal to r_2, is colinear to ℓ_1 (this is where the crucial assumption $N = 2$ comes in!). Multiplying the system to the left by ${}^t\nabla R_2$, we obtain

$$^t\nabla R_2 \partial_t u + \lambda_1 {}^t\nabla R_2 \partial_x u = \partial_t R_2 + \lambda_1 \partial_x R_2 = 0,$$

since $\partial R_2 = {}^t\nabla R_2 \partial u$. Proceeding similarly with R_1, we prove the theorem. $\qquad\qquad\qquad\qquad\qquad\qquad\qquad\qquad\qquad\qquad\qquad\square$

Note the curious crossing of indices. The assumption that K be a diffeomorphism is not unrealistic: in fact, the differential $D_u K$ is a 2×2-matrix with lines respectively proportional to ℓ_2 and ℓ_1; hence $D_u K$ is invertible, and K is at least a local diffeomorphism.

4.6   Shock Curves

For Burgers equation it is easy to construct a shock solution: One can fix arbitrary constant states $u_l \in \mathbf{R}$ and $u_r \in \mathbf{R}$ and separate them by a line $x = \phi(t)$ with $\phi'(t) = \frac{1}{2}(u_l + u_r)$. More generally, one can fix arbitrary C^1 solutions u_l and u_r and separate them by a curve $x = \phi(t)$ solving the differential equation

$$\phi'(t) = \frac{1}{2}(u_l + u_r)(\phi(t), t).$$

To construct shock solutions for a strictly hyperbolic system of conservation laws $\partial_t u + \partial_x(F(u)) = 0$, it is important to notice that the Rankine–Hugoniot relation is a *vector* relation in $\mathbf{R}^N$. We will restrict ourselves to piecewise constant shocks, that is, solutions where two constant states (u_l, u_r) are separated by a line $x = st$. Fixing the constant state $u_l \in \mathbf{R}^N$, we look for $u_r \in \mathbf{R}^N$ and $s \in \mathbf{R}$ such that

$$F(u_r) - F(u_l) = s(u_r - u_l).$$

Theorem 4.12. *Let $u_l \in \mathbf{R}^N$ be given. For each $j = 1, \ldots, N$, there exists, for ϵ close to zero, a C^∞ curve*

$$\epsilon \mapsto (u_r(\epsilon), s(\epsilon)) \in \mathbf{R}^N_u \times \mathbf{R}$$

satisfying

$$F(u_r(\epsilon)) - F(u_l) = s(\epsilon)(u_r(\epsilon) - u_l),$$

and such that

$$u_r(\epsilon) = u_l + \epsilon r_j(u_l) + O(\epsilon^2), \; s(\epsilon) = \lambda_j(u_l) + O(\epsilon).$$

We call this curve the j-shock curve through u_l.

Proof: The problem we have to solve is a *purely algebraic problem*, for which we cannot use the implicit function theorem, since we know (or suspect) that there are N families of solutions. To prepare for the use of the implicit function theorem, we rephrase the problem as follows:

Step 1. Define

$$A(u, v) = \int_0^1 F'(tu + (1-t)v)dt, \; A(u, v) = A(v, u), \; A(u, u) = F'(u),$$

so that

$$F(u) - F(v) = A(u, v)(u - v),$$

and the Rankine–Hugoniot relation looks now like an eigenvalue problem:

$$A(u_r, u_l)(u_r - u_l) = s(u_r - u_l).$$

(If A were a constant matrix, this really would be an eigenvalue problem!) For u and v close to u_l (which we assume from now on), the matrix $A(u, v)$ is close to $A(u_l)$, hence its eigenvalues

$$\lambda_1(u, v) < \cdots < \lambda_N(u, v)$$

are real and distinct. The corresponding right and left eigenvectors are denoted by $r_j(u, v)$ and $\ell_j(u, v)$ as usual. The Rankine–Hugoniot relation can be equivalently rewritten as

$$\forall k, \; [{}^t\ell_k(u_r, u_l)(u_r - u_l)](\lambda_k(u_r, u_l) - s) = 0.$$

We now fix j (recall that u_l is already fixed). Define $\Phi : \mathbf{R}^N \to \mathbf{R}^{N-1}$ by

$$\Phi(u_r) = [{}^t\ell_1(u_r, u_l)(u_r - u_l), \ldots, {}^t\ell_N(u_r, u_l)(u_r - u_l)],$$

where the term with index j has been *omitted*. If u_r satisfies $\Phi(u_r) = 0$, the choice $s = \lambda_j(u_r, u_l)$ provides a solution of the Rankine–Hugoniot relation. We have thus eliminated the unknown s.

Step 2. To solve $\Phi(u_r) = 0$, we use the implicit function theorem: split the variables u_r as

$$u_r = (x, v), \; x = (u_r)_1, \; v = ((u_r)_2, \ldots, (u_r)_N),$$

and think of Φ as a function of (x, v). For $u_r = u_l$, $\Phi = 0$. Assuming for simpicity $j = 1$, we see that the partial derivative $(\partial_v \Phi)(u_l)$ is represented by the $(N - 1) \times (N - 1)$ matrix that is formed by the lines

$$^t\ell_2(u_l, u_l), \ldots, {}^t\ell_N(u_l, u_l),$$

with the first column discarded. Since these lines are independent, some $(N - 1) \times (N - 1)$ minor has to be invertible, we can as well assume that it is $\partial_v \Phi$. Hence we obtain, for x close to $(u_l)_1$, a curve $v = v(x)$ with

$$\Phi(x, v(x)) = 0.$$

Setting $\epsilon = x - (u_l)_1$, and differentiating Φ with respect to ϵ yields, for $\epsilon = 0$,

$$^t\ell_k(u_l, u_l)u_r' = 0, \ k \neq j.$$

Hence u_r' is colinear to $r_j(u_l)$, which gives the theorem. $\qquad\square$

If the eigenvalue λ_j is genuinely nonlinear, we can improve the theorem as follows.

Theorem 4.13. *Assume λ_j to be genuinely nonlinear, and the corresponding r_j normalized. Then the j-shock curve can be parametrized to satisfy*

$$u_r(\epsilon) = u_l + \epsilon r_j(u_l) + O(\epsilon^2), \ s(\epsilon) = \lambda_j(u_l) + \frac{\epsilon}{2} + O(\epsilon^2).$$

Proof: From $s = \lambda_j(u_r(\epsilon), u_l)$, we obtain $s'(0) = \partial_1 \lambda_j(u_l, u_l)u_r'(0)$. On the other hand, since $\lambda_j(u, v) = \lambda_j(v, u)$, then $\partial_1 \lambda_j(u, u) = \partial_2 \lambda_j(u, u)$. Since $\lambda_j(u, u) = \lambda_j(u)$, then $\partial_u \lambda_j(u) = 2\partial_1 \lambda_j(u, u)$. If we choose to parametrize the shock curve in such a way that $u_r'(0)$ is the normalized eigenvector $r_j(u_l)$, we obtain the claim. $\qquad\square$

Corollary 4.14. *Assume the same hypothesis as in Theorem 4.13. Then*

$$s(\epsilon) = \frac{1}{2}[\lambda_j(u_l) + \lambda_j(u_r)] + O(\epsilon).$$

In particular, if $\lambda_j(u_l) \neq \lambda_j(u_r)$, s is different from $\lambda_j(u_l)$ and from $\lambda_j(u_r)$.

This is immediate, since $\lambda_j(u_r) = \lambda_j(u_l) + \epsilon + O(\epsilon^2)$.

4.7   Lax Conditions and Admissible Shocks

In the previous section, we have constructed (small) shock solutions to a strictly hyperbolic system of conservation laws. Which of these solutions are admissible shocks?

Definition 4.15. *A (small amplitude) j-shock solution (u_l, u_r, s) is said to satisfy the Lax conditions if*

$$\lambda_j(u_l) > s > \lambda_j(u_r).$$

We will consider as admissible only the shocks satisfying the Lax conditions. This will be justified by two different types of arguments: a stability argument, which we give below, and an entropy argument, which will be explained in Section 4.10.

In view of Corollary 4.14, an admissible shock for a genuinely nonlinear eigenvalue λ_j is $(u_l, u_r(\epsilon))$ with $\epsilon < 0$: u_r has to be on the corresponding *half*-curve through u_l. Note that, for the same mode j, the (half)-rarefaction curve and the (half)-shock curve through u_l match to form a C^1 curve, representing the states u_r which can be connected to u_l by either a rarefaction wave or a shock wave. We call this curve the *j*-**solution curve through** u_l.

Theorem 4.16. *Let $(\bar u_l, \bar u_r, \bar s)$ be a constant states shock solution to the scalar equation $(N = 1)$*

$$\partial_t u + \partial_x(F(u)) = 0.$$

If this solution is linearly stable, then it satisfies the Lax conditions.

Proof: First, let us explain what "lineary stable" means. Consider the solution $\bar u$ as being $\bar u_l$ for $x < \bar s t$, $\bar u_r$ for $x > \bar s t$. To test stability, the idea is to solve the Cauchy problem for modified initial data $u_l = \bar u_l + \mathring u_l^0$ (for $x < 0$) and $u_r = \bar u_r + \mathring u_r^0$ (for $x > 0$), where $\mathring u_l^0$ and $\mathring u_r^0$ are small. One expects the solution u again to be a shock, with a shock curve $x = \phi(t)$ starting from $(0,0)$, close to $x = \bar s t$, separating states u_r and u_l close to $\bar u_r$ and $\bar u_l$. However, since u_r and u_l are defined in the variable domains $x > \phi(t)$ and $x < \phi(t)$, it is not clear how to define "stability."

To this aim, we transform the problem into a problem where the various unknown functions are defined on *fixed domains*. We perform the change of variables

$$X = x - \phi(t), \ T = t, \ v_r(X,T) = u_r(x,t), \ v_l(X,T) = u_l(x,t).$$

A shock solution $(u_l, u_r, x = \phi(t))$ is tranformed into a solution of the system

$$\partial_T v_l + (F'(v_l) - \phi'(T))\partial_X v_l = 0, \ X < 0,$$
$$\partial_T v_r + (F'(v_r) - \phi'(T))\partial_X v_r = 0, \ X > 0,$$
$$F(v_r) - F(v_l) = \phi'(T)(v_r - v_l), \ X = 0,$$

with the same initial data. The first two equations are the transformed of the equations on u_l and u_r, respectively, while the third is just the Rankine–Hugoniot relation. Suppose that a solution

$$(v_l = \bar{v}_l + \dot{v}_l, \ v_r = \bar{v}_r + \dot{v}_r, \ \phi(T) = \bar{s}T + \dot{\phi})$$

of this new system is closed to the constant states solution ($\bar{v}_l = \bar{u}_l$, $\bar{v}_r = \bar{u}_r$, $\bar{\phi} = \bar{s}T$), that is, $\dot{v}_l, \dot{v}_r, \dot{\phi}$ have magnitude ϵ. We have then

$$\partial_T \dot{v}_l + (F'(\bar{v}_l) - \bar{s})\partial_X \dot{v}_l = q_-, \ X < 0,$$
$$\partial_T \dot{v}_r + (F'(\bar{v}_r) - \bar{s})\partial_X \dot{v}_r = q_+, \ X > 0,$$
$$F'(\bar{v}_r)\dot{v}_r - F'(\bar{v}_l)\dot{v}_l = \dot{\phi}'(\bar{v}_r - \bar{v}_l) + \bar{s}(\dot{v}_r - \dot{v}_l) + q_0, \ X = 0,$$

where the q_+, q_-, and q_0 stand for quantities of magnitude ϵ^2. The system defined by the lefthand sides of the above equations is called the "linearized system on ($\bar{v}_l, \bar{v}_r, \bar{\phi}$)." "Linear stability" means that the linearized system is well-posed, that is, it possesses a unique solution ($\dot{v}_l, \dot{v}_r, \dot{\phi}$) for all righthand sides q_+, q_-, q_0, and initial data ($\dot{v}_l^0, \dot{v}_r^0$).

In the present case, the uniqueness of the solution for the first linearized equation requires

$$F'(\bar{v}_l) - \bar{s} \geq 0,$$

while uniqueness for the second linearized equation requires

$$F'(\bar{v}_r) - \bar{s} \leq 0.$$

To have $\dot{v}_l$ and $\dot{v}_r$ well-defined on $X = 0$ for $T > 0$, we need strict inequalities, and these are exactly the Lax conditions. $\qquad\Box$

4.8   Contact Discontinuities

In dealing with rarefaction waves and shock solutions to strictly hyperbolic systems of conservation laws, we have emphasized the case of genuinely nonlinear eigenvalues. Linearly degenerate eigenvalues, however, exist in natural situations, such as the Euler equations, for example (see Exercise 4). We now briefly discuss this case.

Theorem 4.17. *Let λ_j be a linearly degenerate eigenvalue and let the constant states u_r, u_l belong to the same integral curve of r_j. Then $\lambda_j(u_l) = \lambda_j(u_r) \equiv \lambda$, and the function $u = u_l$ for $x \leq \lambda t$ and $u = u_r$ for $x \geq \lambda t$ is a shock solution.*

Proof: Let $u(s)$ be the integral curve of r_j through the given state u_l: by definition, λ_j is constant along this curve, with value say λ. We claim that $F(u(s)) - \lambda u(s)$ is a constant; In fact, its derivative with respect to s is

$$F'(u(s))r_j(u(s)) - \lambda_j(u(s))r_j(u(s)) \equiv 0.$$

If u_r belongs to the integral curve of r_j through u_l, the constant states function u which is u_l for $x < \lambda t$ and u_r for $x > \lambda t$ is a solution of our system, since

$$F(u_r) - \lambda u_r = F(u_l) - \lambda u_l,$$

which is the Rankine–Hugoniot relation. In other words, the j-solution curve in this case is just the full integral curve of r_j through u_l. □

Note that u can be viewed either as a shock solution for which $\lambda_j(u_l) = \lambda_j(u_r)$ (in sharp contrast with the genuinely nonlinear case!), or as a rarefaction wave for which the fan between $x = s_l t$ and $x = s_r t$ is reduced to one line.

4.9 Riemann Problem

The Riemann problem is the Cauchy problem

$$\partial_t u + \partial_x(F(u)) = 0, \ u(x,0) = u_0(x),$$

where u_0 is a constant state $u_l \in \mathbf{R}^N$ for $x < 0$ and a constant state $u_r \in \mathbf{R}^N$ for $x > 0$. The reasons for studying this particular problem derive from its model character (one expects the solution corresponding to general data with some discontinuity to behave analogously), and from its importance in the theory of finite difference schemes. From Sections 4.4, 4.6, 4.7, and 4.8, we only know how to solve this problem if u_r belongs to some j-solution curve through u_l, λ_j being either genuinely nonlinear or linearly degenerate.

Theorem 4.18. *Assume that all eigenvalues of the system are either genuinely nonlinear or linearly degenerate. Then, for all u_l, there exists $\eta > 0$ such that we can solve the Riemann problem for all initial data (u_l, u_r) with $|u_l - u_r| \leq \eta$.*

Proof: To solve the general Riemann problem, fix u_l and define in a neighborhood V of the origin in $\mathbf{R}^N_\epsilon$ the function

$$V \ni \epsilon \mapsto \Phi(\epsilon_1, \ldots, \epsilon_N) \in \mathbf{R}^N$$

as follows: Let u^1 be the point of parameter ϵ_1 on the 1-solution curve through $u^0 = u_l$, u^2 be the point of parameter ϵ_2 on the 2-solution curve through u^1, and so on until we reach $u^N = \Phi(\epsilon_1, \ldots, \epsilon_N)$. Note that $\Phi(0) =$

u_l. Assume that $D_0\Phi$ is invertible: then Φ is a local diffeomorphism, and all u_r sufficiently close to u_l can be connected to u_l by a broken solution curve corresponding to some ϵ. This broken curve is "coding" for a true solution u of the Riemann problem: u is obtained simply by placing one next to the other, from left to right, the solution patterns corresponding to the situation "u^k is the point of parameter ϵ_k on the k-solution curve through u^{k-1}." For instance, if $N = 3$, all eigenvalues are genuinely nonlinear and $\epsilon_1 < 0, \epsilon_2 > 0, \epsilon_3 < 0$, the solution u is the juxtaposition, from left to right, of a 1-shock separating u_l and u_1, a 2-rarefaction connecting u_1 to u_2, and a 3-shock separating u_2 from u_r.

To prove that $D_0\Phi$ is invertible, we remember that its columns are $(\partial_{\epsilon_1}\Phi(0),\ldots,\partial_{\epsilon_N}\Phi(0))$. Since $\Phi(0,\ldots,0,\epsilon_k,0,\ldots,0)$ is just the point of parameter ϵ_k on the k-solution curve through u_l,

$$\partial_{\epsilon_k}\Phi(0) = r_k(u_l).$$

The eigenvectors r_k being independent, the claim is proved.   □

The case of a linear system (with A a constant matrix) corresponds to all eigenvalues linearly degenerate; all discontinuities are contact discontinuities.

4.10   Viscosity and Entropy

Another approach to admissible solutions of conservation laws is based on physical considerations: the system of conservation laws

$$\partial_t u + \partial_x(F(u)) = 0$$

is viewed as an approximation of a better system, supposedly closer to reality,

$$\partial_t u + \partial_x(F(u)) = \epsilon\partial_x^2 u,$$

as the "viscosity" $\epsilon > 0$ goes to zero. This terminology comes from the case of compressible fluids governed by the complete Euler system with viscosity. One could of course consider that this better system is the only one we should study, but we will not follow this orientation. We ask instead the following question: Suppose there exists a sequence u_ϵ of solutions of the better system that converge to u in such a way that u is a solution to our system of conservation laws. Does u enjoy any special "physical" property to distinguish it from any other solution of the system?

The answer is yes.

4.10.1   Entropy

To display the special features (or at least some special features) of these "physical solutions," we need a new concept.

Definition 4.19. *Let (q, g) be a pair of real C^1 functions on $\Omega \subset \mathbf{R}_u^N$. We say that q is an entropy, with entropy flux g, for the system*

$$\partial_t u + \partial_x(F(u)) = 0$$

if, identically on Ω,

$$^t\nabla q(u)F'(u) = {}^t\nabla g(u).$$

The idea underlying this definition is to find additional *scalar* conservation equations for C^1 solutions of the system. In fact, for all $u \in C^1$ we have the identity

$$\partial_t(q(u)) + \partial_x(g(u)) = [-{}^t\nabla q(u)F'(u) + {}^t\nabla g(u)](\partial_x u) + {}^t\nabla q(u)[\partial_t u + \partial_x(F(u))].$$

For a scalar conservation law $N = 1$, q can be chosen arbitrarily and then g follows by integration. For a 2-system, equality of the cross derivatives of g yields one condition on q, which is a second order PDE on q; If we can find a solution q of this PDE, then g follows (locally) by integration (see Exercise 12). In general, however, q has to satisfy too many equations in order for ${}^tF'(u)\nabla q$ to be the gradient of some function, and there are no such pairs (q, g).

What is the point of the concept if no such objects exist? The point is that for physical systems of interest special symmetries of the system allow q to exist. This is the case for instance for the Euler system, and that is where the name "entropy" comes from (see Exercise 4).

4.10.2   Limits of Viscosity Solutions

The following theorem is the answer to the question asked in the introduction of this section.

Theorem 4.20. *Suppose that there exists a sequence $u_\epsilon \in C^2(\Omega)$ of solutions of*

$$\partial_t u_\epsilon + \partial_x(F(u_\epsilon)) = \epsilon \partial_x^2 u_\epsilon,$$

such that

i) for some M, $|u_\epsilon(x, t)| \leq M$;

ii) for some u, $u_\epsilon \to u$ almost everywhere in Ω.

Then if (q, g) is a pair entropy/entropy flux with q convex,

$$\partial_t(q(u)) + \partial_x(g(u)) \leq 0.$$

This last inequality means that the distribution defined by the lefthand side is in fact a negative measure (this sounds terrible, but we will see below examples of how this works in practice for simple patterns). To prove the theorem, note that, thanks to the Lebesgue dominated convergence theorem,

$$\partial_t(q(u)) + \partial_x(g(u))$$

is the limit, in the sense of distributions, of $Q_\epsilon \equiv \partial_t(q(u_\epsilon)) + \partial_x(g(u_\epsilon))$. By the algebraic relation on (q, g), $Q_\epsilon = \epsilon^t \nabla q(u_\epsilon) \partial_x^2 u_\epsilon$. Now $\partial_x(q(u_\epsilon)) = {}^t \nabla q(u_\epsilon) \partial_x u_\epsilon$,

$$\partial_x^2(q(u_\epsilon)) = {}^t \nabla q(u_\epsilon) \partial_x^2 u_\epsilon + {}^t(\partial_x u_\epsilon) q''(u_\epsilon)(\partial_x u_\epsilon).$$

The convexity of q precisely means that q'' is nonnegative; hence $Q_\epsilon \leq \epsilon \partial_x^2(q(u_\epsilon))$. Since the righthand side above goes to zero with ϵ in the distribution sense, the theorem is proved. $\qquad\qquad\square$

It seems here that some heuristic explanation is needed to really understand the point of this proof: If u is a shock, the solutions u_ϵ realize a smooth transition from u_l to u_r in an interval of size roughly ϵ (see Exercise 14). Hence a derivative of u_ϵ of order k is likely to have magnitude ϵ^{-k}. The terms manipulated in the above proof are thus highly singular in ϵ.

4.10.3   Application to the Case of a Shock

If u is a shock with states u_l and u_r separated by a shock curve $x = \phi(t)$, the same computation as given in Section 4.1 for the proof of the Rankine–Hugoniot relation yields

$$\partial_t(q(u)) + \partial_x(g(u)) = \delta(x - \phi(t))[g(u_r) - g(u_l) - \phi'(q(u_r) - q(u_l))].$$

Hence, in this case, the negative measure condition of Theorem 4.20 simply means

$$g(u_r) - g(u_l) \leq \phi'(q(u_r) - q(u_l)).$$

Let us consider for instance Burgers equation: We can choose

$$q(u) = \frac{u^2}{2}, \ g(u) = \frac{u^3}{3}.$$

Since $\phi' = (u_r + u_l)/2$, the condition reads

$$(u_r - u_l)\left[\frac{u_r^2 + u_r u_l + u_l^2}{3} - \frac{(u_r + u_l)^2}{4}\right] \leq 0,$$

that is $(u_r - u_l)^3/12 \leq 0$, hence $u_r \leq u_l$, which is the Lax condition.

More generally, we have the following theorem.

Theorem 4.21. *Let $\partial_t u + \partial_x(F(u)) = 0$ be a system of conservation laws, λ_j be a genuinely nonlinear eigenvalue, and $u(\epsilon)$ a j-shock curve throught the constant state u_l, namely,*

$$u(\epsilon) = u_l + \epsilon r_j(u_l) + O(\epsilon^2), \ s(\epsilon) = \lambda_j(u_l) + \frac{\epsilon}{2} + O(\epsilon^2).$$

If (q, g) is a pair entropy/entropy flux, then

$$g(u(\epsilon)) - g(u_l) - s(\epsilon)[q(u(\epsilon)) - q(u_l)] = \frac{\epsilon^3}{12}[^t r_j(u_l)q''(u_l)r_j(u_l)] + O(\epsilon^4), \ \epsilon \to 0.$$

If q is strictly convex, admissible shocks for q correspond to $\epsilon < 0$, that is, they satisfy the Lax conditions.

Proof: Set

$$\delta(\epsilon) = g(u(\epsilon)) - g(u_l) - s(\epsilon)[q(u(\epsilon)) - q(u_l)].$$

Taking the derivative with respect to ϵ of the Rankine–Hugoniot relation, we obtain

$$[F'(u(\epsilon)) - s(\epsilon)]u'(\epsilon) = s'(\epsilon)(u(\epsilon) - u_l).$$

Then, using the definition of the pair (q, g) and Taylor formula,

$$\delta'(\epsilon) = {}^t\nabla q(u(\epsilon))[F'(u(\epsilon)) - s(\epsilon)]u'(\epsilon) - s'(\epsilon)[q(u(\epsilon)) - q(u_l)]$$
$$= s'(\epsilon)[q(u_l) - q(u(\epsilon)) - {}^t\nabla q(u(\epsilon))(u_l - u(\epsilon))] = O(\epsilon^2).$$

Hence

$$\delta(0) = \delta'(0) = \delta''(0) = 0, \ \delta'''(0) = \frac{1}{2}{}^t r_j(u_l)q''(u_l)r_j(u_l),$$

which completes the proof.   □

4.11   Exercises

1.(a) Show that if $u \in C^1$ is a real solution of Burgers equation

$$\partial_t u + \partial_x \left(\frac{u^2}{2} \right) = 0,$$

it is also a solution of the equation

$$\partial_t (u^2) + \partial_x \left(\frac{2u^3}{3} \right) = 0.$$

(b) Compare the Rankine–Hugoniot conditions for the two conservation laws. Conclude that two different conservation laws can have the same smooth solutions without having the same shock solutions.

2. For the following systems, compute the eigenvalues and the eigenvectors; discuss conditions for the eigenvalues to be either genuinely nonlinear or linearly degenerate:

(a) The p-system

$$\partial_t u - \partial_x v = 0, \ \partial_t v - \partial_x (p(u)) = 0, \ p' > 0.$$

(b) The isentropic Euler system (in nonconservative form)

$$\partial_t \rho + \partial_x (\rho u) = 0, \ \partial_t u + u \partial_x u + \frac{\partial_x p}{\rho} = 0,$$

where $p \equiv p(\rho)$ is given with $c^2 \equiv dp/d\rho > 0$.

3. For a quasilinear strictly hyperbolic $N \times N$ system

$$\partial_t u + A(u)\partial_x u = 0,$$

the integral curves of $r_j(u)$ contain the images of simple waves. By performing a change of unknowns $u = \Phi(v)$, we obtain a new system

$$\partial_t v + \tilde{A}(v)\partial_x v = 0.$$

Compute its eigenvalues and eigenvectors $\tilde{\lambda}_j(v)$ and $\tilde{r}_j(v)$. Show that the integral curves of r_j are the image by Φ of the integral curves of $\tilde{r}_j$ (check that this is coherent with the way Φ transforms simple waves). Show that the concept of genuinely nonlinear eigenvalue is invariant.

4. Consider the full Euler system from Section 4.1. Show that a C^1 solution (ρ, u, p) of this system is also a solution of the system in nonconservative form

$$D_t\rho + \rho\partial_x u = 0, \quad D_t u + \frac{\partial_x p}{\rho} = 0, \quad D_t e + \frac{p}{\rho}\partial_x u = 0,$$

where $D_t = \partial_t + u\partial_x$. Show that there exists (at least locally in (ρ, p)) a function $s(\rho, p)$ (the "entropy") such that the last equation can be replaced by $D_t s = 0$. Deduce from this that u is a linearly degenerate eigenvalue of the system.

5. Consider $u \in C^2(\mathbf{R} \times [0, T[)$ a solution of the Cauchy problem for the 2-system in diagonal form

$$\partial_t u_1 + \lambda_1(u)\partial_x u_1 = 0, \quad \partial_t u_2 + \lambda_2(u)\partial_x u_2 = 0.$$

Establish the system satisfied by $(\partial_x u_1, \partial_x u_2)$. Determine nonvanishing functions $F_i(u)$ such that, setting $v_i = F_i(u)\partial_x u_i$, one obtains a diagonal system

$$\partial_t v_i + \lambda_i(u)\partial_x v_i + G_i(u)v_i^2 = 0, \quad i = 1, 2.$$

6. Consider a strictly hyperbolic $N \times N$ system of conservation laws

$$\partial_t u + \partial_x(F(u)) = 0.$$

Let $\lambda(u)$ be a real, simple, genuinely nonlinear eigenvalue of $F'(u)$, and $r(u)$ its normalized eigenvector. Fix U_0 and define the integral curve of r

$$U'(s) = r(U(s)), \quad U(0) = U_0.$$

On the other hand, consider a λ-shock curve $(V(t), \mu(t))$ defined by

$$F(V(t)) - F(U_0) = \mu(t)(V(t) - U_0), \quad V(0) = U_0, \quad \mu(0) = \lambda(U_0).$$

Show that one can find a function $\phi(s)$, with $\phi(0) = 0$, $\phi'(0)$, and $\phi''(0)$ appropriately chosen such that

$$U(s) - V(\phi(s)) = O(s^3).$$

Deduce from this that the solution curve for λ is indeed C^2.

7. Consider Burgers equation with a "source term"

$$\partial_t u + u\partial_x u = u^2.$$

(a) Let C be a real constant: Compute explicitly the solution u_C with initial value C. Compute the integral curves γ of the vector field $\partial_t + u_C \partial_x$.

(b) Let now $C_- < C_+$ be two given real constants. Let $\gamma_\pm$ be the integral curves through the origin of the fields $\partial_t + u_{C_\pm} \partial_x$. In the sector between γ_- and γ_+, we set

$$u(x,t) = \frac{e^x - 1}{t}.$$

Check that u is indeed a solution of the equation in this sector, matching on $\gamma_\pm$ the solutions $u_{C_\pm}$ so as to form a continuous solution of the equation for $t > 0$ small enough.

8. Consider the p-system

$$\partial_t u = \partial_x v, \ \partial_t v = \partial_x(p(u)),$$

where we assume $p' > 0, p'' > 0$.

(a) Write the Rankine–Hugoniot relation for a shock connecting the constant state (u_r, v_r) (on the right) to the constant state (u_l, v_l) (on the left), separated by a line of speed s. Considering the left state as given, determine explicitly the shock curves, taking $\epsilon = u_r - u_l$ as a parameter. Write down the Lax conditions as inequalities on u_r and u_l.

(b) Compute explicitly the Riemann invariants for this system. Explain how this system can be diagonalized.

9. Consider a strictly hyperbolic quasilinear $N \times N$ system

$$\partial_t u + A(u)\partial_x u = 0.$$

(a) Let u be a k-simple wave. Show that u is constant along the integral curves of $\partial_t + \lambda_k(u)\partial_x$. Deduce from this that these curves are in fact straight lines.

(b) In the case of a diagonal 2-system, show that a C^1 solution u, constant with value C to the right of an integral curve γ of $\partial_t + \lambda_2(u)\partial_x$ is 2-simple to the left of γ. What can be said for a general 2-system?

10. Consider a strictly hyperbolic 2-system of conservation laws with genuinely nonlinear eigenvalues, say

$$\lambda_1(u) < 0 < \lambda_2(u).$$

Suppose given a solution for $t \leq 0$ which consists of three constant states (u_-, v, u_+) (from left to right), separated by straight lines $x = s^2_- t$ and $x = s^1_- t$ (this solution represents the collision at the origin at time zero of a 1-shock and a 2-shock, moving with respective speeds s_-^1 and s_-^2 towards each other). We assume $|u_+ - u_-| = \epsilon$ small enough, and we want to compute the solution for positive values of time.

(a) Draw a picture in the plane $\mathbf{R}_u^2$ of

i) the solution curves connecting v to u_- (for the value ϵ_2 of the parameter) and u_+ to v (for the value ϵ_1),

ii) the solution curves corresponding to the solution of the Riemann problem for $t \geq 0$ and initial values u_- (left) and u_+ (right).

Show that the solution of (ii) consists of a 1-shock with speed s_+^1 connecting w to u_- (with parameter ϵ_1') and a 2-shock with speed s_+^2 connecting u_+ to w (with parameter ϵ_2'). Show also that

$$\epsilon_1' = \epsilon_1 + O(\epsilon^2), \quad \epsilon_2' = \epsilon_2 + O(\epsilon^2).$$

(b) Compute, modulo $O(\epsilon^2)$, the changes of speed $s_+^i - s_-^i, i = 1, 2$ (Note that, in the linear case, the speeds do not change!)

(c) Using the same method, study the interaction of a rarefaction wave coming from the left with a shock coming from the right; do the same for two rarefaction waves.

11. Consider the p-system

$$\partial_t u = \partial_x v, \quad \partial_t v = \partial_x(u^3/3).$$

We want to solve the Riemann problem with (constant) initial data $(u_l > 1, v_l)$ for $x < 0$ and (u_r, v_r) for $x > 0$.

(a) We take first

$$\bar{u}_r = u_l - 1, \quad \bar{v}_r = v_l - u_l + \frac{1}{2}.$$

Show that the solution is a 1-rarefaction wave.

(b) We take now

$$u_r = u_l - 1, \quad v_r = v_l - u_l + \frac{1}{2} + \eta = \bar{v}_r + \eta.$$

Assuming η small enough, discuss according to the sign of η the nature of the solution. Compute modulo $O(\eta^2)$ the magnitude of the new wave.

12. Consider the p-system

$$\partial_t u = \partial_x v, \ \partial_t v = \partial_x(p(u)), \ p' > 0.$$

Find the necessary and sufficient second order PDE q has to satisfy in order to be an entropy function. Show that this equation is hyperbolic. What are its characteristic fields? How can one construct convex entropies?

13. Let q be a convex entropy for a system

$$\partial_t u + \partial_x(F(u)) = 0.$$

Show that q'' is a symmetrizer for the linearized system $\partial_t + F'(u)\partial_x$, that is, $q''F'$ is symmetric.

14. Consider a scalar equation "with viscosity"

$$\partial_t u + \partial_x(F(u)) = \epsilon \partial_x^2 u, \ F'' > 0, \ \epsilon > 0.$$

We look for a *formal* solution of this equation, i.e.

$$u_\epsilon(x,t) = \Sigma_{k \geq 0} \epsilon^k U_k(s,t), \ s = \frac{x - \phi(t)}{\epsilon},$$

where the functions U_k have limits $U_k^{\pm}$ independent of t as $s \to \pm\infty$. Such a solution would represent a quick smooth transition between the states $\Sigma \epsilon^k U_k^-$ (left) and $\Sigma \epsilon^k U_k^+$ (right). Formally, the functions u_ϵ converge to the function u which is U_0^- left of $x = \phi(t)$ and U_0^+ to the right. Hence we assume that (U_0^-, U_0^+, ϕ') satisfy the Rankine–Hugoniot relation.

Establish the equation satisfied by U_0, and integrate it once using the function

$$\Phi(u) = F(u) - F(U_0^-) - \phi'(t)(u - U_0^-).$$

We remark that the Rankine–Hugoniot relation is just $\Phi(U_0^+) = 0$. Show that the existence of an appropriate U_0 implies the Lax conditions

$$F'(U_0^-) \geq \phi' \geq F'(U_0^+).$$

4.12　Notes

In this chapter, we concentrated on constant state solutions and simple waves, our aim being to present the main concepts (shock and rarefaction waves, entropy) and explain how to solve the Riemann problem. This seems to be the minimal knowledge required to understand the constructions of finite difference schemes in numerical analysis or further theoretical developments. A simple account of conservation laws from this point of view can be found in Chapters 15–20 of Smoller [21]. Simple waves are used intensively in the superb book by Courant and Friedrichs [7], which offers a rich view of many physical problems. Some more information about blowup of smooth solutions is contained in Alinhac [4]. The full mathematical theory of solutions of conservation laws can be found in Hörmander [10] or in Serre [19]. The book by Majda [17] concentrates on multidimensional issues.

Chapter 5

The Wave Equation

In this chapter, we review quickly the main properties of the solutions of
the wave equation

$$\Box \equiv \partial_t^2 - \Delta_x, \ \Delta_x = \Delta = \partial_{x_1}^2 + \cdots + \partial_{x_n}^2$$

in $\mathbf{R}_x^n \times \mathbf{R}_t$, concentrating on the cases $n = 2$ and $n = 3$. Since we promised
not to use distribution theory, we will make no attempt to prove the solution
formulas in the most general context. It is understood that the functions we
manipulate are supposed to allow the formula to be defined in the classical
sense.

5.1   Explicit Solutions

To analyze the Cauchy problem for the wave equation

$$\Box u = f, \ u(x,0) = u_0(x), \ (\partial_t u)(x,0) = u_1(x),$$

we concentrate first on the homogeneous case $f \equiv 0$, with $u_0 = 0$, and
denote the corresponding solution by $u = S(u_1)$. Then, the solution for
$f = 0$ and general (u_0, u_1) is

$$u = S(u_1) + \partial_t S(u_0),$$

since, for $t = 0$, $\partial_t^2 S(u_0) = \Delta_x S(u_0) = 0$.

S. Alinhac, *Hyperbolic Partial Differential Equations*, Universitext,
DOI 10.1007/978-0-387-87823-2_5, © Springer Science+Business Media, LLC 2009

5.1.1   Fourier Analysis

The simplest way to analyze the Cauchy problem for the wave equation is to perform a Fourier transformation with respect to x alone; With an obvious abuse of notation, define

$$\hat{u}(\xi, t) = \int_{\mathbf{R}^n} e^{-ix\xi} u(x, t) dx.$$

Then, formally, the Cauchy problem becomes

$$\partial_t^2 \hat{u}(\xi, t) + |\xi|^2 \hat{u}(\xi, t) = \hat{f}(\xi, t), \ \hat{u}(\xi, 0) = \hat{u}_0(\xi), \ (\partial_t \hat{u})(\xi, 0) = \hat{u}_1(\xi).$$

This is a Cauchy problem for an ODE in t, which can be solved explicitly.

Theorem 5.1. *Assume that u_0, u_1, and f decay enough as $|x| \to +\infty$. Then the solution u of the Cauchy problem with data (u_0, u_1, f) is given by the formula*

$$\hat{u}(\xi, t) = \hat{u}_0(\xi) \cos t|\xi| + \hat{u}_1(\xi) \frac{\sin t|\xi|}{|\xi|} + \int_0^t \hat{f}(\xi, s) \frac{\sin(t-s)|\xi|}{|\xi|} ds.$$

To see this, we look for $\hat{u}$ (according to the method of variation of parameters) in the form

$$\hat{u}(\xi, t) = A(\xi, t) \cos t|\xi| + B(\xi, t) \sin t|\xi|.$$

We impose then

$$\partial_t A \cos t|\xi| + \partial_t B \sin t|\xi| = 0, \ -\partial_t A \sin t|\xi| + \partial_t B \cos t|\xi| = \frac{\hat{f}}{|\xi|},$$

and the initial conditions $A(\xi, 0) = \hat{u}_0(\xi), B(\xi, 0) = \hat{u}_1(\xi)/|\xi|$. This gives the formula.   □

In particular, $\hat{S}(v)(\xi, t) = \hat{v}(\xi) \sin t|\xi|/|\xi|$, and we can check directly the formula given at the beginning of this section. We also remark the fact known as the *Duhamel principle*.

Principle 5.2 (Duhamel principle) *The solution u of the Cauchy problem with zero initial data and righthand side f is given by*

$$u(x, t) = \int_0^t S(f(\cdot, s))(x, t-s) ds.$$

5.1.2   Solutions as Spherical Means

Because of Duhamel principle, it is enough to obtain a formula for $S(v)$, the solution of the Cauchy problem with $f \equiv 0$ and data $(0, v)$.

Theorem 5.3. *In the case $n = 3$,*

$$S(v)(x, t) = \frac{t}{4\pi} \int_{|y|=1} v(x - ty) d\sigma_1(y) = \frac{1}{4\pi t} \int_{|y|=t} v(x - y) d\sigma_t(y).$$

Here, $d\sigma_R$ is the surface element on the sphere of radius R.

Proof: Taking into account the formula

$$\hat{S}(v)(\xi, t) = \hat{v}(\xi) \frac{\sin t |\xi|}{|\xi|},$$

it is enough to prove

$$(\widehat{d\sigma_R})(\xi) = 4\pi R \frac{\sin R|\xi|}{|\xi|},$$

since then $(v * \widehat{d\sigma_t})(\xi) = 4\pi t \hat{v}(\xi) \sin t |\xi|/|\xi| = 4\pi t \hat{S}(v)(\xi, t)$. We use spherical coordinates (for details, see Section 5.2)

$$x_1 = R \sin \phi \cos \theta, \ x_2 = R \sin \phi \sin \theta, \ x_3 = R \cos \phi, \ d\sigma_R = R^2 \sin \phi d\phi d\theta.$$

Since $d\sigma_R$ is invariant by rotations, so is its Fourier transform, so we can take

$$\xi = |\xi|(0, 0, 1).$$

For this value of ξ, we can compute explicitly the integral

$$\widehat{d\sigma_R}(\xi) = \int e^{-ix_3|\xi|} R^2 \sin \phi d\phi d\theta = 2\pi R^2 \int e^{-iR|\xi|u} du = 4\pi R \frac{\sin R|\xi|}{|\xi|}.$$

This is the formula of the theorem. $\square$

It is important to notice that $S(v)(x, t)$ is t **times the spherical mean** of the function v on the sphere of center x and radius t.

To handle the case $n = 2$, we use the so-called "method of descent" introduced by Hadamard.

Theorem 5.4. *In the case $n = 2$,*

$$S(v)(x, t) = \frac{1}{2\pi} \int_{|y| \leq t} v(x - y)(t^2 - |y|^2)^{-1/2} dy.$$

Proof: We can think of v as a function on $\mathbf{R}^3$ independent of x_3: then, since the solution is unique, Sv is also independent of x_3, and

$$S(v)(x_1, x_2, t) = \frac{1}{4\pi t} \int_{|y|=t} v(x_1 - y_1, x_2 - y_2) d\sigma_t(y).$$

When parametrizing the (half)-sphere of radius t in $\mathbf{R}_y^3$ by (y_1, y_2), we have

$$d\sigma_t(y) = t(t^2 - y_1^2 - y_2^2)^{-1/2} dy_1 dy_2,$$

and this gives the result.   □

5.1.3   Finite Speed of Propagation, Domains of Determination

Inspection of the above solution formula shows the two following important facts.

Theorem 5.5. *For $n = 3$, the value of the solution $S(v)$ at (x, t) depends only on the values of v on the sphere of radius t centered at x. For $n = 2$, the value of the solution $S(v)$ at (x, t) depends only on the values of v on the ball of radius t centered at x.*

In other words, the propagation speed is at most one in all directions. What happens to v at points y further away from x than t is not seen by $S(v)(x, t)$; the information, leaving y at time $t = 0$, has not yet reached x at time t. Hence, the information speed is less than one.

More generally, for $n = 3$, the solution $u(x, t)$ of the full Cauchy problem depends on the values of u_0, u_1, and $\partial_n u_0$ (the normal derivative) on the sphere of radius t centered at x, and on the values of f on the lateral boundary of the truncated cone

$$C_{(x,t)} = \{(y, s), \ 0 \le s \le t, \ |y - x| \le t - s\}.$$

In the case $n = 2$, the sphere has to be replaced by the ball, and the boundary of the truncated cone by the full truncated cone. We can now introduce a new concept, similar to the concept introduced for systems in the plane in Definition 2.18.

Definition 5.6. *Let D be a closed domain in $\mathbf{R}_x^n \times [0, \infty[$ with base $\omega = \{(x, 0) \in D\}$. We say that D is a **domain of determination** of* *its base ω if, for all $(x,t) \in D$, the full truncated cone $C_{(x,t)}$ is contained in D.*

For instance, for all $M > 0$ and $z \geq M$, the "spherical" region

$$\{(x,t), \ t \geq 0, \ (t+z)^2 + |x|^2 \leq z^2 + M^2\}$$

is a domain of determination of its base $\{|x| \leq M\}$. The following statement, which explains the name "determination," is an immediate corollary of the solution formula.

Corollary 5.7. *Let D be a domain of determination of its base ω in $\{t = 0\}$. If $u \in C^2(D)$ is a solution of the Cauchy problem in D with $f = 0$ in D and $u_0, u_1 = 0$ in ω, then $u = 0$ in D.*

5.1.4 Strong Huygens Principle

In the case $n = 3$, the solution formula display a remarkable phenomenon, which is called *lacuna*: $S(v)(x,t)$ depends only on the values of v on the *sphere* of radius t centered at x. What happens to v at points y closer to x than t has already been seen by $S(v)(x,t)$. For instance, suppose that u is a solution of the homogeneous Cauchy problem ($f \equiv 0$) with data u_0 and u_1 vanishing for $|x| \geq M$. Then

$$(t \geq M, \ |x| \leq t - M) \Rightarrow u(x,t) = 0.$$

Hence, in this case, not only $u(x,t) = 0$ for $|x| \geq t + M$, which is a consequence of the propagation with speed one, but also $u(x,t) = 0$ inside an upside down cone with vertex at $(0, M)$. This property, specific of the case $n = 3$, will be used in Section 6.6.

5.1.5 Asymptotic Behavior of Solutions

When the Cauchy data u_0, u_1 of a solution u are C_0^∞ functions, the solution formula allows a precise description of this solution for large time.

Theorem 5.8. *Consider, for $n = 3$, the solution $u = S(v)$ of the Cauchy problem*

$$\Box u = 0, \ u(x,0) = 0, \ (\partial_t u)(x,0) = v(x),$$

where $v \in C^\infty$ vanishes for $|x| \geq M$. Then, using polar coordinates $r = |x|$, $x = r\omega$, we have for $t \geq 2M$ the representation

$$u(x,t) = \frac{1}{r} F\left(r - t, \ \omega, \ \frac{1}{r}\right),$$

where

$$F : \mathbf{R} \times S^2 \times \left[0, \frac{1}{M}\right] \ni (\rho, \omega, z) \mapsto F(\rho, \omega, z) \in \mathbf{C}$$

is a C^∞ function, vanishing for $|\rho| \geq M$.

Proof: Set $\rho = r - t$ and take for simplicity $x = r(1, 0, 0)$. From Sections 5.1.3 and 5.1.4 we already know that u vanishes for $|\rho| \geq M$. For $|\rho| \leq M$ and big t (and r), the portion of the sphere of radius t centered at x touching the support of v is very small. It can be parametrized by $y \in \mathbf{R}^2$ as $(r - (t^2 - |y|^2)^{1/2}, y)$. The surface element $d\sigma_t$ (very close to dy for large t) is given by $d\sigma_t = t(t^2 - |y|^2)^{-1/2} dy$. Hence

$$S(v)(x, t) = (4\pi)^{-1} \int_{|y| \leq M} v(r - (t^2 - |y|^2)^{1/2}, y)(t^2 - |y|^2)^{-1/2} dy.$$

We write now

$$(t^2 - |y|^2)^{1/2} = r \left[\left(1 - \frac{\rho}{r}\right)^2 - \frac{|y|^2}{r^2}\right]^{1/2},$$

which shows that $r(t^2 - |y|^2)^{-1/2}$ and

$$r - (t^2 - |y|^2)^{1/2} = \left[1 + \left(\left(1 - \frac{\rho}{r}\right)^2 - \frac{|y|^2}{r^2}\right)^{1/2}\right]^{-1} \left[2\rho + \frac{|y|^2 - \rho^2}{r}\right]$$

are C^∞ functions of $(\rho, 1/r, y)$. If we had taken $x = r\omega$ to start with, we would have obtained C^∞ functions of ω as well. This proves the theorem. $\qquad\qquad\square$

It is important (especially when dealing with nonlinear perturbations of the wave equation) to notice the "decay rate" t^{-1} of the solution.

If we define the "profile at infinity" as $F_0(\rho, \omega) = F(\rho, \omega, 0)$, we observe that

$$4\pi F_0(\rho, \omega) = \int_{\omega.y = \rho} v(y) dy$$

is the *Radon transform* of v.

We leave as an exercise for the reader the corresponding statement for $n = 2$ (Exercise 4).

5.2   Geometry of The Wave Equation

In analyzing the behavior of the solutions of the wave equation, it is convenient to distinguish special vector fields, the *Lorentz vector fields*, which are the usual rotations, the hyperbolic rotations, and the scaling operator. We explain here their definitions and properties.

5.2.1   Rotations and Scaling Vector Field

In this section we discuss rotation and scaling vector fields in $\mathbf{R}_x^2$ or $\mathbf{R}_x^3$ without any reference to the wave equations, to which we return in Section 5.2.2.

Rotation and scaling fields in $\mathbf{R}_{x,y}^2$. In the plane $\mathbf{R}_{x,y}^2$ define the usual polar coordinates by

$$x = r\cos\theta, \ y = r\sin\theta, \ r = (x^2 + y^2)^{1/2}, \ \omega = (\cos\theta, \sin\theta).$$

For $u \in C^1(\mathbf{R}^2)$, setting $v(r, \theta) = u(r\cos\theta, r\sin\theta)$, we get

$$r\partial_r v(r, \theta) = (x\partial_x u + y\partial_y u)(r\cos\theta, r\sin\theta),$$
$$\partial_\theta v(r, \theta) = (x\partial_y u - y\partial_x u)(r\cos\theta, r\sin\theta).$$

We write this formulas abusively

$$r\partial_r = x\partial_x + y\partial_y, \partial_\theta = x\partial_y - y\partial_x.$$

The vector field $S = x\partial_x + y\partial_y$ is the **scaling vector field**, and its integral curve starting from the point m_0 is the ray through the origin $t \mapsto e^t m_0$. We also note that if f is a C^1 positively homogeneous function of degree μ, that is

$$f(\lambda x, \lambda y) = \lambda^\mu f(x, y), \ \lambda > 0,$$

the **Euler identity** (obtained by differentiating the above formula with respect to λ) reads

$$(Sf)(x, y) = \mu f(x, y).$$

The vector field $R = x\partial_y - y\partial_x$ is the **rotation field**, since its integral curve starting from $m_0 = (x_0, y_0)$ is the circle centered at the origin

$$t \mapsto (x_0\cos t - y_0\sin t, x_0\sin t + y_0\cos t).$$

Note that the vectors

$$(\cos\theta, \sin\theta), \ (-\sin\theta, \cos\theta)$$

form an orthonormal basis of the plane; we say that

$$\partial_r = \cos\theta\partial_x + \sin\theta\partial_y, \ r^{-1}\partial_\theta = -\sin\theta\partial_x + \cos\theta\partial_y,$$

form, at each point (x, y), an **orthonormal frame**. The decomposition of the usual vector fields ∂_x and ∂_y on this frame is given by the formula

$$\partial_x = (\cos\theta)\partial_r - (\sin\theta)(r^{-1}\partial_\theta), \ \partial_y = (\sin\theta)\partial_r + (\cos\theta)(r^{-1}\partial_\theta).$$

Finally, straightforward computations give the following commutations formula

$$[R, \partial_r] \equiv R\partial_r - \partial_r R = 0, \ [R, \Delta] \equiv R\Delta - \Delta R = 0, \ [S, \Delta] = -2\Delta,$$

where $\Delta = \partial_x^2 + \partial_y^2$ is the Laplace operator in the plane. This Laplacian is expressed in polar coordinates by

$$\Delta = \partial_r^2 + r^{-1}\partial_r + r^{-2}\partial_\theta^2,$$

since

$$\partial_\theta^2 = y^2\partial_x^2 + x^2\partial_y^2 - 2xy\partial_{xy}^2 - r\partial_r,$$
$$r^2(\partial_r^2 + r^{-1}\partial_r) = (r\partial_r)^2 = x^2\partial_x^2 + y^2\partial_y^2 + 2xy\partial_{xy}^2 + r\partial_r.$$

Rotations and scaling fields in $\mathbf{R}_x^3$. In the space $\mathbf{R}_x^3$, the polar coordinates are

$$x = r\omega, \ r = |x|, \ \omega \in S^2.$$

For ω on the unit sphere S^2, we use the spherical coordinates (θ, ϕ), where θ is the longitude with respect to the plane $\{x_2 = 0\}$ and ϕ the (co)latitude with respect to the x_3-axis. Thus

$$x_1 = r\sin\phi\cos\theta, \ x_2 = r\sin\phi\sin\theta, \ x_3 = r\cos\phi.$$

For $u \in C^1(\mathbf{R}^3)$, setting $v(r, \theta, \phi) = u(r\sin\phi\cos\theta, r\sin\phi\sin\theta, r\cos\phi)$, we obtain $r\partial_r v = \Sigma x_i\partial_i u$, and

$$\partial_\theta v = x_1\partial_2 u - x_2\partial_1 u, \ \partial_\phi v = \frac{\cos\phi}{\sin\phi}(x_1\partial_1 u + x_2\partial_2 u) - (x_1^2 + x_2^2)^{1/2}\partial_3 u.$$

As in the previous discussion, the vector field $S = \Sigma x_i\partial_i$ is called the **scaling vector field**, and its integral curves are the rays through the origin.

We define now the three **rotation fields** R_1, R_2, R_3 by $R = x \wedge \partial$, that is

$$R_1 = x_2\partial_3 - x_3\partial_2, \ R_2 = x_3\partial_1 - x_1\partial_3, \ R_3 = x_1\partial_2 - x_2\partial_1.$$

Since $\nabla r = x/r$, we see that $R_i r = 0$, which means that, for any point m, the fields $R_i(m)$ are tangent to the sphere through m centered at the origin. Since we have three (independent) vector fields tangent to the 2-submanifold S^2, there must be some relation between them; this relation is clearly $\Sigma x_i R_i = 0$. By inspection, we also obtain the formula

$$\partial_\theta = R_3, \partial_\phi = -\sin\theta R_1 + \cos\theta R_2.$$

At any point m not on the x_3-axis, the three vector fields

$$\partial_r, \ e_\theta = (r\sin\phi)^{-1}\partial_\theta, \ e_\phi = r^{-1}\partial_\phi$$

form an **orthonormal frame**. The decomposition of the usual vector fields ∂_i on this frame is given by the formula

$$\partial = \omega\partial_r - \omega \wedge \left(\frac{R}{r}\right).$$

This formula can be checked by straightforward computation, for instance

$$\left(\omega \wedge \frac{R}{r}\right)_1 = \omega_2\frac{R_3}{r} - \omega_3\frac{R_2}{r} = \omega_2(\omega_1\partial_2 - \omega_2\partial_1) - \omega_3(\omega_3\partial_1 - \omega_1\partial_3) = -\partial_1 + \omega_1\partial_r.$$

It is sometimes handy to use the notation $\bar\partial_i = \partial_i - (x_i/r)\partial_r$. These vector fields are tangent to the spheres (this follows from the previous formula, or from the direct observation that $\bar\partial_i r = 0$); for any real function $u \in C^1(\mathbf{R}^3)$, we have the decomposition formulas

$$\Sigma(\partial_i u)^2 = (\partial_r u)^2 + \Sigma(\bar\partial_i u)^2 = (\partial_r u)^2 + \left|\frac{R}{r}u\right|^2.$$

As in dimension $n = 2$, there is also the remarkable commutation formula:

$$[R_i, \partial_r] = 0, \ [R_i, \Delta] = 0, \ [S, \Delta] = -2\Delta,$$

where $\Delta = \Sigma\partial_i^2$ is the usual Laplace operator. For the first formula, we have, for instance,

$$r[R_1, \partial_r] = [R_1, \Sigma x_i \partial_i] = R_1 - R_1 = 0.$$

For the second we get from the definition $[R_1, \partial_1^2] = 0$,

$$[R_1, \partial_2^2]u = x_2\partial_3\partial_2^2 u - x_3\partial_2^3 u - \partial_2^2(x_2\partial_3 u - x_3\partial_2 u) = -2\partial_{23}^2 u,$$

and similarly $[R_1, \partial_3^2]u = 2\partial_{23}^2 u$, which gives the result. In the same way, we obtain

$$[S, \partial_i^2]u = \Sigma x_j\partial_j\partial_i^2 u - \partial_i^2(\Sigma x_j\partial_j u) = -2\partial_i^2,$$

which gives the last formula.

To express the Laplacian in polar coordinates, we first compute

$$\Delta_\omega \equiv \Sigma R_i^2 = (x_2^2 + x_3^2)\partial_1^2 + (x_1^2 + x_3^2)\partial_2^2 + (x_1^2 + x_2^2)\partial_3^2$$
$$- 2\Sigma_{i<j} x_i x_j \partial_{ij}^2 - 2r\partial_r,$$
$$r^2(\partial_r^2 + 2r\partial_r) = (r\partial_r)^2 + r\partial_r = \Sigma x_i \partial_i^2 + 2\Sigma_{i<j} x_i x_j \partial_{ij}^2 + 2r\partial_r,$$

to obtain finally

$$\Delta = \partial_r^2 + \frac{2}{r}\partial_r + r^{-2}\Delta_\omega.$$

The operator

$$\Delta_\omega = \Sigma R_i^2$$

is called the "Laplace operator on the unit sphere." A tedious computation shows that

$$\Delta_\omega = \partial_\phi^2 + [(\sin\phi)^{-1}\partial_\theta]^2 + \frac{\cos\phi}{\sin\phi}\partial_\phi.$$

5.2.2   Hyperbolic Rotations and Lorentz Fields

We return now to the space time $\mathbf{R}_x^n \times \mathbf{R}_t$ for $n = 2$ or $n = 3$, and to the wave equation $\Box = \partial_t^2 - \Delta$. In analogy with the usual spatial rotations $R = x \wedge \partial$, we define the **hyperbolic rotations** H_i by

$$H_i = t\partial_i + x_i\partial_t, \ i = 1,\dots,n.$$

These vector fields are tangent to the hyperboloids $\{t^2 - |x|^2 = C\}$, which explains their names. We have seen that the spatial rotations commute with $\Box$, $[\Box, R] = 0$. Just the same, the fundamental property of the hyperbolic rotations is the commutation relations

$$[H_i, \Box] = 0.$$

This is easily established since

$$[t\partial_i, \partial_t^2] = -2\partial_{ti}^2, \ [t\partial_i, \partial_j^2] = 0, \ [x_i\partial_t, \partial_t^2] = 0, \ [x_i\partial_t, \partial_i^2] = -2\partial_{ti}^2,$$

and $[x_i\partial_t, \partial_j^2] = 0$ for $j \neq i$. We define the **Lorentz vector fields** to be

i) the usual derivatives ∂_t, ∂_i $(i = 1,\dots,n)$;

ii) the spatial rotations $R = x_1\partial_2 - x_2\partial_1$ (for $n = 2$) or $R = x \wedge \partial$ (for $n = 3$);

iii) the hyperbolic rotations $H_i = t\partial_i + x_i\partial_t$ $(i = 1,\dots,n)$;

iv) the scaling vector field $S = t\partial_t + \Sigma x_i \partial_i$.

In the sequence the letter Z will denote any one of these vector fields. The following theorem summarizes the commutation relations that we have obtained.

Theorem 5.9. *All Lorentz vector fields commute with the wave operator, except for S, which satisfies*

$$[\Box, S] = 2\Box.$$

5.2.3   Light Cones, Null Frames, and Good Derivatives

We saw in section 5.1.3 the importance of the backwards cone $C_{x,t}$ for the wave equation. It turns out that the forward cones $\{t - r = C\}$ also play an important role. We restrict our attention here to the case $n = 3$, though analogous and simpler considerations hold for the case $n = 2$. Let us define first the two (orthogonal) vector fields

$$L = \partial_t + \partial_r, \ \underline{L} = \partial_t - \partial_r.$$

At any point, the vector fields

$$(L, \underline{L}, e_\theta, e_\phi)$$

form an orthonormal basis which is called a **null frame** (see Exercise 9 to understand the origin of this terminology). For any point $m_0 = (x_0, t_0)$, $x_0 \neq 0$, the tangent space at m_0 to the forward cone $\{t - r = t_0 - r_0\}$ through m_0 is spanned by the three vectors

$$L, \ e_\theta, \ e_\phi,$$

or, equivalently,

$$r^{-1} R_1, \ r^{-1} R_2, \ r^{-1} R_3, \ L = \partial_t + \partial_r$$

at this point.

The reason for introducing these fields is best seen by considering a solution $u = S(v)$ with $v \in C^\infty$ vanishing for $|x| \geq M$: we have the representation formula (see 5.1.5)

$$u(x,t) = r^{-1} F\left(r - t, \omega, \frac{1}{r}\right),$$

where F is a C^∞ function vanishing for $|\rho = r - t| \geq M$. We establish easily

$$\underline{L}u = -2r^{-1}(\partial_\rho F)\left(r - t, \omega, \frac{1}{r}\right) + O(r^{-2}), \ Lu = O(r^{-2}), \ \left(\frac{R}{r}\right)u = O(r^{-2}).$$

Thus, as far as the behavior of u when $t \to +\infty$ is concerned, $\underline{L}u$ behaves like t^{-1}, while the other derivatives Lu and $(R/r)u$ behave like t^{-2}. In this context, we will speak of "bad" or "good" derivatives.

The difference between "bad" and "good" derivatives can also be seen by relating them to the Lorentz fields. In fact, we have the formula

$$\Sigma \omega_i H_i = t\partial_r + r\partial_t, \ S = t\partial_t + r\partial_r,$$
$$(r + t)L = \Sigma \omega_i H_i + S, \ (r - t)\underline{L} = \Sigma \omega_i H_i - S,$$
$$(t + r)\frac{R}{r} = R + \omega \wedge H.$$

Using these formulas, the following theorem is easily proved.

Theorem 5.10. *There exists a constant C such that, for all $u \in C^1$ ($\mathbf{R}_x^n \times \mathbf{R}_t$) and $t \geq 0$,*

$$(1 + r + t)|(Lu)(x,t)| \leq C(\Sigma|Zu|)(x,t),$$
$$(1 + r + t)|r^{-1}(R_i u)(x,t)| \leq C(\Sigma|Zu|)(x,t),$$
$$(1 + |r - t|)|(\underline{L}u)(x,t)| \leq C(\Sigma|Zu|)(x,t),$$

where the sums are extended over all Lorentz vector fields. In particular,

$$(1 + |r - t|)|(\partial u)(x,t)| \leq C(\Sigma|Zu|)(x,t),$$

where ∂ stands for ∂_t or ∂_i.

The importance of these formulas appears when one studies the behavior at infinity of global solutions of the wave equation. The key fact is that the Lorentz fields Z commute nicely with $\Box$, while the fields L, R/r do not. Suppose that, for a certain C^1 function u, and all Lorentz fields Z, we have obtained a control

$$|Zu|(x,t) \leq M(x,t)$$

by a certain function M, say, going to zero at infinity. This implies, thanks to the above formula, for $t \geq 0$ and irrelevant numerical constant C,

$$(1 + r + t)|(Lu)(x,t)| \leq CM(x,t),$$
$$(1 + r + t)|r^{-1}(R_i u)(x,t)| \leq CM(x,t),$$
$$(1 + |r - t|)|(\underline{L}u)(x,t)| \leq CM(x,t).$$

In other words, the "good" derivatives $(Lu, r^{-1}R_i u)$ decay faster than M at infinity by a factor $1 + r + t$, while the "bad" derivative $\underline{L}u$ decays faster only away from the light cone with equation $\{t = r\}$.

Let us say a few words here to explain why we emphasize the role played by the Lorentz fields: What we have in mind is to obtain the qualitative behavior of global solutions to wave equations with variable coefficients (see Chapter 7), or to nonlinear wave equations. Let P be a given second order operator close to $\Box$: the Lorentz fields Z do not commute any more with the given operator P, but the commutator $[P, Z]$ may have small coefficients, so that, writing

$$PZu = Z(Pu) + [P, Z]u,$$

it is possible to get a control $|Zu| \leq M(x, t)$ by a known function as above. The point of the argument is that it does not rely on an explicit representation formula, but on a priori inequalities for P (see Chapter 7 for more details).

5.2.4   Klainerman's Inequality

We will see in subsequent chapters that it is essential in the study of hyperbolic multidimensional equations or systems to use L^2 norms. To connect these norms with L^∞ norms, a simple tool is the so-called "Sobolev Lemma."

Lemma 5.11 (Sobolev Lemma). *For all positive integers m, n, with $m > n/2$, there exist constants $C_{n,m}$ such that for all $u \in S(\mathbf{R}_x^n)$ (the Schwartz space),*

$$\|u\|_{L^\infty} \leq C_{n,m} \Sigma_{|\alpha| \leq m} \|\partial_x^\alpha u\|_{L^2}.$$

Proof: Since the Fourier transform of $\partial_x^\alpha u$ is $i^{|\alpha|} \xi^\alpha \hat{u}(\xi)$, by Plancherel's theorem,

$$A^2 \equiv \Sigma_{|\alpha| \leq m} |\partial_x^\alpha u|_{L^2}^2 \geq C \int (1 + \Sigma \xi_i^{2m}) |\hat{u}(\xi)|^2 d\xi.$$

We write now $\hat{u}$ as

$$\hat{u} = [\hat{u}(1 + |\xi|^{2m})^{1/2}][(1 + |\xi|^{2m})^{-1/2}].$$

Since $m > n/2$, the function in the last bracket belongs to L^2; since, for some $C > 0$,

$$1 + |\xi|^{2m} \leq C(1 + \Sigma \xi_i^{2m}),$$

the L^2 norm of the function in the first bracket is less than CA. By the
Fourier inversion formula and Hölder inequality, we thus obtain

$$||u||_{L^\infty} \leq C||\hat{u}||_{L^1} \leq CA,$$

which is the inequality we wanted to prove.   □

The Klainerman inequality is best understood when one compares it to
the Sobolev lemma: In this lemma, we use L^2 norms of products ∂_x^α of
the usual derivatives applied to u to obtain a control of $|u(x,t)|$. In the
Klainerman inequality, the idea is to use, in $\mathbf{R}_x^3 \times \mathbf{R}_t$, L^2 norms of products
Z^k of *Lorentz vector fields* Z applied to u to obtain a *weighted* control of
$|u(x,t)|$.

Theorem 5.12 (Klainerman inequality) *There is a constant C such
that, for all $u \in \mathcal{S}(\mathbf{R}_x^n \times [t-1, t+1])$,*

$$(1 + |t| + r)^{n-1}(1 + ||t| - r|)|u(x,t)|^2 \leq C\Sigma_{0 \leq k \leq (n+2)/2}|Z^k u(\cdot, t)|_{L^2}^2.$$

*Here Z^k in the righthand side means a product of k of the Lorentz vector
fields Z, and the sum is extended to all such products.*

Proof: We will not give a complete proof of this inequality (see
Hörmander's book). We restrict ourselves to $n = 3$ and prove only

$$(1 + t + r)^2|u(x,t)|^2 \leq C\Sigma_{0 \leq k \leq 2}|Z^k u(\cdot, t)|_{L^2}^2$$

for $t/2 \leq r \leq 3t/2$, $t \geq 1$. In this region $\mathcal{R}$, for fixed t, we use the variables
$\rho = r - t, \theta, \phi$. We have

$$\partial_\rho = \partial_r = \Sigma\omega_i\partial_i, \ \partial_\theta = R_3, \ \partial_\phi = -\sin\theta R_1 + \cos\theta R_2,$$

hence derivatives like $\partial_\rho^k\partial_\phi^p\partial_\theta^q u$ are bounded by a constant times $\Sigma|Z^m u|$,
where $m \leq k + p + q$ and we use in fact only the fields $Z = \partial_i$ or $Z = R$.
Thus

$$\Sigma_{k+p+q \leq 2}\int_{|\rho| \leq t/2}|\partial_\rho^k\partial_\phi^p\partial_\theta^q u|^2 dx \leq C\Sigma_{m \leq 2}|Z^m u(\cdot, t)|_{L^2}^2.$$

But $dx = (t + \rho)^2 d\rho \sin\phi d\theta d\phi$, so that, writing the lefthand side as an
integral in (ρ, θ, ϕ), we gain a factor t^2. We use now a slight extension of
Sobolev lemma *in these variables* (ρ, θ, ϕ), saying that for some constant C
(independent of t),

$$||u||_{L^\infty(R)} \leq C\Sigma_{k+p+q \leq 2}\int_R|\partial_\rho^k\partial_\phi^p\partial_\theta^q u|^2 d\rho d\phi d\theta,$$

and this yields the result.   □

We see how this inequality can be combined with the inequalities of
Theorem 5.10 to give a rather good description of the behavior of
solutions.

5.3 Exercises

1.(a) For $n = 3$, prove the formula

$$\Box\left(\frac{v}{r}\right) = \left(\frac{1}{r}\right)[\partial_t^2 - \partial_r^2 - r^{-2}\Delta_\omega]v.$$

Analogously, compute $\Box(v/\sqrt{r})$ for $n = 2$ and compare. In the sequence we always assume $n = 3$.

(b) Suppose that the function $u(r, t)$ can be written

$$u(r, t) = z(r^2, t)$$

for some $z \in C^\infty$. Then u is thought of as an even function of r defined for all $r \in \mathbf{R}$. Show that, if such a u is a solution of the wave equation, $v = ru$ satisfies on $\mathbf{R}_r \times \mathbf{R}_t$

$$\partial_t^2 v - \partial_r^2 v = 0.$$

Write then explicitly u knowing its Cauchy data $(u_0(r), u_1(r))$. As an application, compute the solution u with Cauchy data $(r^2, 0)$.

(c) More generally, if $u(r, t)$ is a rotationally invariant solution of $\Box u = 0$, write explicitly u knowing its Cauchy data. As an application, compute the solution u with Cauchy data $(r, 0)$.

(d) Compute explicitly u with data $(0, v(r))$ when v is the characteristic function of the ball of radius a. Show that u is discontinuous at $(0, a)$.

2.(a) For $n = 3$, consider the solution $u = S(v)$ and its explicit expression as a spherical mean of v. Using Stokes formula (see Appendix, Formula A.17), transform the integral of v on the sphere of center x and radius t into the integral on the ball of the divergence of some field (Hint: One can write

$$v(y) = \Sigma v(y)\left[\frac{y_i - x_i}{t}\right]\left[\frac{y_i - x_i}{t}\right].$$

(b) Assuming $|v| + |\nabla v| \in L^1(\mathbf{R}_x^3)$, deduce from (a) that, for some $C \geq 0$ and $t \geq 1$,

$$|u(x, t)| \leq \frac{C}{t}.$$

3. Let $\lambda \in \mathbf{C}$ be a given number, and let $v \in C^2(\mathbf{R}_x^2 \times \mathbf{R}_t)$ be the solution of the Cauchy problem

$$\partial_t^2 v - (\partial_1^2 + \partial_2^2)v + \lambda^2 v = 0, \quad v(x, 0) = v_0(x), \quad (\partial_t v)(x, 0) = v_1(x).$$

Prove that
$$u(x_1, x_2, x_3, t) = e^{i\lambda x_3} v(x_1, x_2, t)$$
is a solution of the wave equation in $\mathbf{R}_x^3 \times \mathbf{R}_t$. Deduce from this a representation formula for v.

4. For $n = 2$ prove that, if $v \in C_0^\infty$, the solution $S(v)$ has the polar coordinate representation
$$u(x, t) = r^{-1/2} F\left(r - t, \omega, \frac{1}{r}\right)$$
for a C^∞ bounded function F.

5. Consider, in $\mathbf{R}_x^n \times \mathbf{R}_t$ ($n = 2$ or $n = 3$), C^∞ solutions u and v of
$$\Box u = f, \ \Box v = g$$
with zero Cauchy data. Prove that if $|f| \leq g$, then $|u| \leq v$.

6. In $\mathbf{R}_x^3$, establish that $[R_i, R_j] = -\epsilon_{ijk} R_k$, where ϵ_{ijk} is the signature of the permutation
$$(1, 2, 3) \mapsto (i, j, k).$$

Compute $[S, R_i]$. Is it possible to give a geometric interpretation of the result using Exercise 14 of Chapter 1?

7. In $\mathbf{R}_x^3 \times \mathbf{R}_t$, away from $\{r = 0\}$, define the vector fields
$$T_i = \partial_i + \omega_i \partial_t, \ \omega = \frac{x}{r}.$$

Prove the formula
$$T_i = \bar{\partial}_i + \omega_i L, \ \omega \wedge T = \frac{R}{r}$$
and show that, at each point m_0, the fields T_i span the tangent space at m_0 to the cone of equation $\{t - r = C\}$ through m_0.

8. Define on $\mathbf{R}_x^3 \times \mathbf{R}_t$ the functions
$$u(x, t) = r + t, \ \underline{u}(x, t) = t - r.$$

Show that they are solutions (for $r \neq 0$) of the eikonal equation
$$(\partial_t \phi)^2 - \Sigma(\partial_i \phi)^2 = 0.$$

Establish the relations
$$2\partial_u = L, \ 2\partial_{\underline{u}} = \underline{L}, \ 2S = uL - \underline{u}\underline{L}.$$

9.(a) In the interior of the light cone

$$\Omega = \{(x,t) \in \mathbf{R}^3_x \times \mathbf{R}_t, r < |t|\},$$

we define the conformal inversion I by the formula

$$I(x,t) = (X,T), \quad X = (t^2 - r^2)^{-1}x, \quad T = -(t^2 - r^2)^{-1}t.$$

Show that I maps Ω into itself, and that $I^2 = id$.

(b) Suppose that u and v are two C^1 functions defined on Ω with $u = v(I)$. Show then

$$(\partial_i u)(x,t) = [(T^2 - R^2)\partial_{X_i} v + 2X_i(T\partial_T v + R\partial_R v)](X,T), \quad R = |X|,$$
$$(\partial_t u)(x,t) = [(R^2 + T^2)\partial_T v + 2RT\partial_R v](X,T),$$

and find the transformed of the Lorentz vector fields R_i, H_i, and S. Show also that L and $\underline{L}$ transform respectively to

$$(T + R)^2(\partial_T + \partial_R), \quad (T - R)^2(\partial_T - \partial_R).$$

The vector field
$$K_0 = (r^2 + t^2)\partial_t + 2rt\partial_r,$$

which appears as the transform of ∂_T by I, will be important in Chapter 7. Show that it can be written as

$$K_0 = \frac{1}{2}[(t + r)^2 L + (t - r)^2 \underline{L}] = u^2 \partial_u + \underline{u}^2 \partial_{\underline{u}}.$$

(c) On $\mathbf{R}^3_x \times \mathbf{R}_t$, we define the (nonpositive) scalar product of two vectors

$$U = (U_1, U_2, U_3, U_0), \quad V = (V_1, V_2, V_3, V_0),$$

by the formula

$$< U, V >= U_0 V_0 - U_1 V_1 - U_2 V_2 - U_3 V_3.$$

Show that

$$< L, L >= 0, \quad < \underline{L}, \underline{L} >= 0, \quad < L, \underline{L} >= 2, \quad < L, r^{-1} R_i >= 0.$$

In the vocabulary of relativity theory, the scalar product $< \cdot >$ is the metric, and the basis

$$(L, \underline{L}, e_\theta, e_\phi)$$

is a null frame. Show that for any point $m = (x, t) \in \Omega$ and any vector V, the vector $W = (D_m I)(V)$ satifies

$$< W, W > = (t^2 - r^2)^{-2} < V, V >.$$

This is the reason why I is said to be "conformal."

(d) In the case $n = 3$, using the above computations and the formula

$$\Box \left(\frac{w}{r} \right) = \left(\frac{1}{r} \right) (L\underline{L} - r^{-2} \Delta_\omega) w,$$

prove that $\Box u = f$ implies

$$(\partial_T^2 - \Delta_X) \left(\frac{v}{T^2 - R^2} \right) = (T^2 - R^2)^{-3} f(I), \quad v = u(I).$$

What is the corresponding formula for $n = 2$?

(e) Consider two functions $u_0, u_1 \in C^\infty(\mathbf{R}_x^3)$ vanishing for $|x| \geq M$. Fix some $t_0 > M$, and let u be the solution of the Cauchy problem

$$\Box u = 0, \quad u(x, t_0) = u_0(x), \quad (\partial_t u)(x, t_0) = u_1(x),$$

for $t \geq t_0$. We know that u is supported in the set $r \leq t - t_0 + M$. Prove that the traces $v(X, T_0)$ and $(\partial_T v)(X, T_0)$ of the transformed function $v = u(I)$ on the plane $\{T = T_0 = -1/t_0\}$ are C^∞ functions v_0 and v_1, compactly supported in the set $|X| < -T_0$. Recover from (d) the representation formula for u obtained in Section 5.1.5.

10. Let $v \in C_0^\infty(\mathbf{R}_x^2)$, and $u = Sv$. Using the formula from Theorem 5.4, show that, for $|x| \leq \alpha(1 + t)$, $\alpha < 1$,

$$|u(x, t)| \leq \frac{C}{1 + t}, \quad |\partial u(x, t)| \leq \frac{C}{(1 + t)^2}.$$

5.4   Notes

Basic informations about the wave equation can of course be found in all PDE books, for instance Evans [9], Hörmander [10], John [12], Lax [15], Taylor [23]. A simple account (with exercises) in the framework of distribution theory is Zuily [25]. We put emphasis on the geometry of the wave equation, reflecting recent research progress; this material is taken mainly from Hörmander [10] or Shatah-Struwe [20]. One can also read Christodoulou and Klainerman [6].

Chapter 6

Energy Inequalities for the Wave Equation

We explain in this chapter what energy inequalities for the wave equation are, and how to obtain them, starting from the simplest cases.

6.1   Standard Inequality in a Strip

We define the strip $S_T \subset \mathbf{R}_x^n \times \mathbf{R}_t$ to be

$$S_T = \{(x,t),\ x \in \mathbf{R}^n,\ 0 \le t < T\}.$$

Let $u \in C^2(S_T)$ be a solution of the Cauchy problem

$$\Box u \equiv (\partial_t^2 - \Delta_x)u = f,\ u(x,0) = u_0(x),\ (\partial_t u)(x,0) = u_1(x).$$

Assumptions 6.1:

  i) the function u is real;

  ii) for each t, $0 \le t < T$, $u(\cdot,t)$ has compact support.

To obtain an energy inequality, we always proceed in three steps.

Step 1. Establishing a differential identity. Using the simple formula

$$(\partial_t^2 u)(\partial_t u) = \frac{1}{2}\partial_t(\partial_t u)^2,$$

S. Alinhac, *Hyperbolic Partial Differential Equations*, Universitext,
DOI 10.1007/978-0-387-87823-2_6, © Springer Science+Business Media, LLC 2009

$$(\partial_i^2 u)(\partial_t u) = \partial_i[(\partial_i u)(\partial_t u)] - (\partial_i u)(\partial_{ti}^2 u) = \partial_i[(\partial_i u)(\partial_t u)] - \frac{1}{2}\partial_t[(\partial_i u)^2],$$

we write the product $(\Box u)(\partial_t u)$ in *divergence form*

$$(\Box u)(\partial_t u) = \frac{1}{2}\partial_t[\Sigma(\partial_i u)^2 + (\partial_t u)^2] - \Sigma\partial_i[(\partial_i u)(\partial_t u)].$$

Note that due to the magic of integration by parts, the minus sign before Δ in the wave operator has changed to a plus sign in the expression $|\partial u|^2 \equiv \Sigma(\partial_i u)^2 + (\partial_t u)^2$.

Step 2. Integration over the domain. Using the above expression in divergence form, we integrate over the strip S_t for $t < T$:

$$\int_{S_t} (\Box u)(\partial_t u)dxds = E(t) - E(0),$$

where the **energy of** u at time t is defined by

$$E(t) = E_u(t) = \frac{1}{2}\int [\Sigma(\partial_i u)^2 + (\partial_t u)^2](x,t)dx = \frac{1}{2}\int |\partial u|^2(x,t)dx.$$

We have obtained the following theorem.

Theorem 6.2. *If $f \equiv 0$, the energy of the solution u is constant in time.*

In the inhomogeneous case $f \not\equiv 0$, we proceed further.

Step 3. Handling the remainder term

To handle the term $\int_{S_t}(\Box u)(\partial_t u)dxds$, we proceed in two steps:

(a) First, we apply the Cauchy–Schwarz inequality for fixed s:

$$\left|\int_0^t ds \int (\Box u)(\partial_t u)dx\right| \le \int_0^t ds\|f(\cdot,s)\|_{L^2}\|(\partial_t u)(\cdot,s)\|_{L^2}$$

$$\le \int_0^t ds\|f(\cdot,s)\|_{L^2}(2E(s))^{1/2}.$$

(b) Next, we introduce the quantity $\phi(t) = \max_{0\le s\le t} E(s)^{1/2}$. For $0 \le t' \le t < T$, by extending the domain of integration, we get

$$E(t') \le E(0) + \int_0^t ds\|f(\cdot,s)\|_{L^2}(2E(s))^{1/2},$$

hence, by taking the supremum in t',

$$\phi(t)^2 \le \phi(0)^2 + \sqrt{2}\phi(t)\int_0^t \|f(\cdot,s)\|_{L^2}ds.$$

We have proved the following theorem.

Theorem 6.3. *Let $u \in C^2(S_T)$ satisfy the Assumptions 6.1. For $t < T$, the following a priori energy inequality holds*

$$\max_{0 \leq s \leq t} E(s)^{1/2} \leq E(0)^{1/2} + \sqrt{2} \int_0^t \|(\Box u)(\cdot, s)\|_{L^2} ds.$$

We explain now how to extend Theorem 6.3 to more general functions u than that satisfying Assumptions 6.1.

First, to deal with the case of a complex u, it is possible, of course, to split u into real and imaginary part

$$u = v + iw, \quad \Box u = g + ih, \quad \Box v = g, \quad \Box w = h.$$

Writing separately the inequalities for v and w yields an inequality for u. A more elegant way to do this is as follows: In step 1, we consider the quantity $(\Box u)(\partial_t \bar{u})$ instead of $(\Box u)(\partial_t u)$. Then

$$(\partial_t^2 u)(\partial_t \bar{u}) = \partial_t(|\partial_t u|^2) - (\partial_t u)(\partial_t^2 \bar{u}),$$

which yields $2\Re[(\partial_t^2 u)(\partial_t \bar{u})] = \partial_t(|\partial_t u|^2)$. Similarly,

$$(\partial_i^2 u)(\partial_t \bar{u}) = \partial_i[(\partial_i u)(\partial_t \bar{u})] - (\partial_i u)(\partial_{ti}^2 \bar{u}), \quad 2\Re[(\partial_i u)(\partial_{ti}^2 \bar{u})] = \partial_t(|\partial_i u|^2).$$

Gathering the terms, we obtain finally the new differential identity of step 1:

$$2\Re[(\Box u)(\partial_t \bar{u})] = \partial_t(\Sigma|\partial_i u|^2 + |\partial_t u|^2) - \Sigma \partial_i\{2\Re[(\partial_i u)(\partial_t \bar{u})]\}.$$

The energy of u at time t is now defined by

$$E(t) = E_u(t) = \frac{1}{2} \int [\Sigma|\partial_i u|^2 + |\partial_t u|^2](x, t)dx.$$

Steps 2 and 3 are exactly the same as before, hence both theorems remain true for complex functions u. In the next sections, for simplicity, we prove all inequalities *only for real* u, leaving the extension to complex u to the reader.

The assumption 6.1(ii) about u is satisfied if (u_0, u_1) vanish for $|x| \geq M$ and $\Box u$ vanishes for $|x| \geq M + t$. However, all we need in the proof of Theorem 6.3 is

$$\int_{\mathbf{R}^n} \Sigma \partial_i[(\partial_i u)(\partial_t u)]dx = 0$$

for the terms appearing in the differential identity of step 1 of the proof. This will be the case also if u is sufficiently decaying when $|x| \rightarrow +\infty$ for fixed $t < T$. In the sequence we will always use this expression in the statements. The question about telling from the data (f, u_0, u_1) whether the solution u of the Cauchy problem is sufficiently decaying when $|x| \rightarrow +\infty$ is touched upon in Exercise 3.

6.2   Improved Standard Inequality

We called the energy inequality in Section 6.1 "standard inequality." It turns out that a somewhat more careful computation gives additional control of some special derivatives.

Theorem 6.4. *For all $\epsilon > 0$, there exists a constant C_ϵ such that, for all $u \in C^2(\mathbf{R}_x^n \times [0, \infty[)$ sufficiently decaying when $|x| \rightarrow +\infty$,*

$$\max_{0 \leq s \leq t} E(s)^{1/2} + \left[\int_{S_t} (1 + |r - s|)^{-1-\epsilon} [\Sigma(T_i u)^2](x, s) dx ds \right]^{1/2}$$

$$\leq C_\epsilon \left[E(0)^{1/2} + \int_0^t ||(\Box u)(\cdot, s)||_{L^2} ds \right].$$

Here,

$$r = |x|, \ \omega = x/r, \ (T_i u)(x, t) = [\partial_i u + \omega_i \partial_t u](x, t).$$

Proof: The proof of the theorem is a typical example of the proof of a *weighted inequality*. It follows the same essential Steps 1, 2, and 3 as in Section 6.1.

Step 1. Establishing a differential identity. For a smooth function $a(x, t)$ to be chosen later, we write

$$(\Box u)(\partial_t u)e^a = \frac{1}{2}\partial_t[e^a(\Sigma(\partial_i u)^2 + (\partial_t u)^2)] - \Sigma\partial_i[e^a(\partial_i u)(\partial_t u)] + \frac{e^a}{2}Q,$$

where

$$Q = -(\partial_t a)[\Sigma(\partial_i u)^2 + (\partial_t u)^2] + 2(\partial_t u)\Sigma(\partial_i a)(\partial_i u).$$

This is an extension of the differential identity "in divergence form" previously obtained: We have now terms in divergence form *plus* a sum Q of quadratic terms in $\partial_i u, \partial_t u$ (with coefficients depending on $\partial_i a, \partial_t a$).

We choose now $a(x,t) = b(r - t)$. The additional quadratic terms in the above expression become then

$$Q = b'[\Sigma(\partial_i u)^2 + (\partial_t u)^2 + 2(\partial_t u)(\partial_r u)] = b'\Sigma(T_i u)^2.$$

If we choose $b'(s) = (1 + |s|)^{-1-\epsilon}$, the function b is bounded, and so is a.

Step 2. Integration over the domain. We now integrate in the strip S_t to obtain, with $\Box u = f$,

$$E^a(t) - E^a(0) + \frac{1}{2} \int_{S_t} e^a(1 + |r - s|)^{-1-\epsilon}[\Sigma(T_i u)^2](x, s) dx ds$$

$$\leq \int_{S_t} e^a |f||\partial_t u| dx ds,$$

where the *modified* energy of u at time t is

$$E^a(t) = \frac{1}{2} \int e^a[\Sigma(\partial_i u)^2 + (\partial_t u)^2](x, t) dx.$$

Note that the integrand in the expression of E^a is the same as before, only multiplied by e^a. Since a is bounded, there exists $c_\epsilon > 0$ such that, for all t,

$$c_\epsilon^{-1} E(t) \leq E^a(t) \leq c_\epsilon E(t).$$

All the exponential factors can be cancelled from the inequality and replaced by appropriate constants. The end of the proof (Step 3) is then exactly the same as before. □

It is important to understand what we really gain in the above theorem: in a region

$$\{|r - t| \geq C_0 t\}, \ C_0 > 0,$$

(that is, deep inside the light cone $\{r = t\}$ or far away from it), the factor $(1 + |r - t|)^{-1-\epsilon}$ is less than $C(1 + t)^{-1-\epsilon}$, hence integrable. Thus, for some constant C_ϵ and any t,

$$\int_{S_t \cap \{|r-s| \geq C_0 s\}} (1 + |r - s|)^{-1-\epsilon}[\Sigma(T_i u)^2](x, s) dx ds \leq C_\epsilon \max_{0 \leq s \leq t} E(s).$$

This is an already controlled quantity, and there is no improvement over the standard inequality in this region. In contrast, in a thin strip, say $|r - t| \leq C_0$, around the light cone $\{r = t\}$, we obtain that the *special energy*

$$\tilde{E}(t) = \frac{1}{2} \int [\Sigma(T_i u)^2](x, t) dx$$

is not just bounded, but also an L^2 function of t.

To understand how this can be possible, we remark first that the special derivatives T_i are the "good" derivatives in the sense of Section 5.2.3, since $L = \partial_t + \partial_r = \Sigma \omega_i T_i, R/r = \omega \wedge T$, and also $T_i = \bar{\partial}_i + \omega_i L = [(R/r) \wedge \omega]_i + \omega_i L$. In the case $u = S(v)$ with $v \in C_0^\infty$, the representation formula of Theorem 5.8 for u shows that all derivatives of u are less than C/t, while the special derivatives $T_i u$ satisfy $|T_i u| \leq C/t^2$.

6.3   Inequalities in a Domain

Consider a domain $D \subset \mathbf{R}_x^n \times \mathbf{R}_t$ which is an open subset of the closed half-space

$$\{(x,t),\ x \in \mathbf{R}^n,\ t \geq 0\}.$$

For $t \geq 0$, we define $\Sigma_t = \{x, (x,t) \in D\}, D_t = \{(x,s) \in D, 0 \leq s \leq t\}$. Thus the "base" of $\bar{D}$ is $\bar{\Sigma}_0$, D_t consists of the points of D between Σ_0 and Σ_t, and we denote by Λ_t the "lateral boundary" of D_t,

$$\Lambda_t = \{(x,s) \in \partial \bar{D}_t,\ s \neq 0,\ s \neq t\}.$$

We assume that on the upper part of $\partial \bar{D}$ there exists piecewise a unit outward normal N to ∂D, namely,

$$N = (N_1, \ldots, N_n, N_0),\ N_0 > 0.$$

Let u be a real function in $C^2(\bar{D})$ with traces $u_0(x) = u(x,0)$ and $u_1(x) = (\partial_t u)(x,0)$ on $\bar{\Sigma}_0$, and $\Box u = f$. We define the **energy of** u at time t by

$$E(t) = E_u(t) = \frac{1}{2} \int_{\Sigma_t} [(\Sigma(\partial_i u)^2 + (\partial_t u)^2](x,t)dx.$$

Theorem 6.5. *Assume that the unit outward normal N to the upper part of ∂D satisfies*

$$N_0^2 \geq \Sigma N_i^2.$$

Then, for all $u \in C^2(\bar{D})$ sufficiently decaying when $|x| \to +\infty$,

$$\max_{0 \leq s \leq t} E(s)^{1/2} \leq E(0)^{1/2} + \sqrt{2} \int_0^t \|f(\cdot,s)\|_{L^2(\Sigma_s)}ds.$$

Proof: The proof follows closely the proof of Theorem 6.3. Here, step 1 is exactly the same as in section 6.1. We write

$$(\Box u)(\partial_t u) = \frac{1}{2}\partial_t[\Sigma(\partial_i u)^2 + (\partial_t u)^2] - \Sigma \partial_i[(\partial_i u)(\partial_t u)].$$

For the integration over D_t in step 2, we use Stokes formula, which we recall here (see also Appendix, Formual A.17).

Formula 6.6 (Stokes formula). *Let* $X = (X_1, \ldots, X_p)$ *be a vector field on a domain* $\Omega \subset \mathbf{R}^p$, *which possesses piecewise a unit outward normal* N *on its boundary* $\partial\Omega$. *Then*

$$\int_\Omega (\Sigma \partial_i X_i)(x)dx = \int_{\partial\Omega} (X \cdot N)d\sigma,$$

where $d\sigma$ *is the area element on* $\partial\Omega$.

Using formula 6.6 we integrate $(\Box u)(\partial_t u)$ in D_t and obtain

$$\int_{D_t} (\Box u)(\partial_t u)dxds = E(t) - E(0) + \int_{\Lambda_t} I d\sigma,$$

with

$$2I = N_0[\Sigma(\partial_i u)^2 + (\partial_t u)^2] - 2(\partial_t u)\Sigma N_i \partial_i u.$$

Forming squares, we obtain

$$2I = N_0\Sigma(\partial_i u - (N_i/N_0)\partial_t u)^2 + N_0^{-1}(\partial_t u)^2(N_0^2 - \Sigma N_i^2).$$

The integral term on $\int_{D_t}((\Box u)(\partial_t u)dxds$ is handled exactly as in step 3 above. $\qquad\square$

In general, the following terminology is used for vectors $N = (N_1, \ldots, N_n, N_0)$:

i) N is "timelike" if $N_0^2 > \Sigma N_i^2$;

ii) N is "null" if $N_0^2 = \Sigma N_i^2$;

iii) N is "spacelike" if $N_0^2 < \Sigma N_i^2$.

The condition on N in Theorem 6.5 is that N is nonspacelike, and this is equivalent to asking that ∂D has slope at most 1 everywhere. From Chapter 2 we know that this is a necessary condition for uniqueness when $n = 1$. Simple geometric considerations show that this condition on N is equivalent to saying that D is a domain of determination of its base Σ_0 for the wave equation, in the sense of Chapter 5. The remarkable fact is that this necessary assumption on D is also sufficient to make our method work (it could be that this method would require stronger assumptions).

It is important to notice that the boundary terms on Λ_t, which we have discarded in the proof of the above theorem, may have some interest. For instance, assume that D is a (truncated) light cone. In this case, the outgoing normal N is a null vector. The vector fields $\partial_i - (N_i/N_0)\partial_t$ span the tangent space to Λ_t, and the term

$$2\int_{\Lambda_t} I d\sigma = \int_{\Lambda_t} N_0\{\Sigma[\partial_i u - (N_i/N_0)\partial_t u]^2\}d\sigma$$

gives the additional control of the L^2 norm on Λ_t of those derivatives of u which are *tangent* to Λ_t.

6.4   General Multipliers

Consider again a domain D as in Section 6.3 and $u \in C^2(\bar{D})$. Up to now, to establish an energy inequality, we have considered the quantity $\int_{D_t}(\Box u)(\partial_t u)dx ds$. More generally, one can consider the quantity $\int_{D_t}(\Box u)$ $(Xu)dx ds$, where $X = X_0\partial_t + \Sigma X_i\partial_i$ (with $X_0 > 0$) is a given vector field, called the *multiplier*. The first step in the proof can be carried out as usual.

Step 1. Establishing a differential identity. The following identity holds

$$(\Box u)(Xu) = \frac{1}{2}\partial_t\{X_0[\Sigma(\partial_i u)^2 + (\partial_t u)^2] + 2(\partial_t u)\Sigma X_i\partial_i u\}$$
$$+ \frac{1}{2}\Sigma\partial_i\{X_i[\Sigma(\partial_i u)^2 - (\partial_t u)^2] - 2X_0(\partial_t u)(\partial_i u)$$
$$- 2(\partial_i u)\Sigma X_j\partial_j u\} + Q.$$

Here, Q denotes a quadratic form in the derivatives of u whose coefficients are derivatives of the components X_0, X_i, of X, which we do not attempt to write explicitly.

Step 2. Integration over the domain. When integrating $(\Box u)(\partial_t u)$ on D_t with outer normal N, we obtain from the divergence terms in $(\Box u)(\partial_t u)$ the boundary terms $\int_{\partial D_t} e d\sigma$ where the **energy density** $e = e(N, X)$ is

$$e = \frac{1}{2}(N_0 X_0 - \Sigma N_i X_i)(\partial_t u)^2 + \frac{1}{2}(N_0 X_0 + \Sigma N_i X_i)\Sigma(\partial_i u)^2$$
$$+ (\partial_t u)(N_0 A - X_0 B) - AB,$$

with

$$A = \Sigma X_i\partial_i u = Xu, \ B = \Sigma N_i\partial_i u = Nu.$$

The integral

$$E_u(t) = \int_{\Sigma_t} e d\sigma$$

is called the energy of u at time t, and we have

$$\int_{\partial D_t} e d\sigma = E_u(t) - E_u(0) + \int_{\Lambda_t} e d\sigma.$$

i) In the special case where $N = (0,1)$ (the flat part of ∂D_t), we obtain

$$e = X_0[\Sigma(\partial_i u)^2 + (\partial_t u)^2] + 2(\partial_t u)A,$$

$$e = X_0 \Sigma[\partial_i u + \left(\frac{X_i}{X_0}\right)\partial_t u]^2 + X_0^{-1}(\partial_t u)^2(X_0^2 - \Sigma X_i^2).$$

Hence, if X is nonspacelike, the quadratic form e is nonnegative, and so is the energy.

ii) On the other hand, if $X = \partial_t$, on the lateral boundary Λ_t of D_t,

$$e = N_0[\Sigma(\partial_i u)^2 + (\partial_t u)^2] - 2(\partial_t u)B,$$

and we proved in Theorem 6.5 that it is nonnegative if N is nonspacelike.

The following theorem summarizes the interplay between the choice of X and the geometry of ∂D.

Theorem 6.7. *Assume that, on the upper part of ∂D, N and X are nonspacelike. Then*

$$\int_{D_t} (\Box u)(Xu)dxds = E_u(t) - E_u(0) + \int_{\Lambda_t} e d\sigma + \int_{D_t} Q dxds,$$

where $E_u(t)$ and e are nonnegative.

The proof of the general case is a tedious but straightforward computation that we leave to the reader. One can also find in Exercise 13 a more geometric proof.   □

This theorem leaves of course open the question about the sign of Q, which is a delicate question in general. We saw in Section 6.2 a first example, where $X = e^{b(r-t)}\partial_t$ generates a nonnegative Q. The next section is devoted to a second example where $X = \partial_r$, generating an "almost nonnegative" Q, and a third example will be seen in Section 6.7 where $X = (r^2+t^2)\partial_t+2tr\partial_r$ also generates an "almost nonnegative" Q. Other examples are given in Exercise 4 and Exercises 7, 8, 10, 12 of Chapter 7.

6.5   Morawetz Inequality

In constrast with Sections 6.1-6.4 where we use timelike multipliers to obtain energy inequalities, the computation below due to Morawetz uses the spacelike multiplier $X = \partial_r$. Of course, in this case, the energy on horizontal slabs $\{t = T\}$ will not be positive, but interesting quadratic terms Q in ∂u will appear. This inequality can only be used when coupled with an energy inequality giving a control of $E_u(t)$.

Theorem 6.8. *Let $u \in C^2(\mathbf{R}_x^3 \times [0, T[)$ be sufficiently decaying when $|x| \to \infty$ and satisfy $\Box u = 0$. Then for all $0 \leq t < T$,*

$$4\pi \int_0^t u^2(0, s)ds + \int_{S_t} r^{-1}[\Sigma(\partial_i u)^2 - (\partial_r u)^2]dxds \leq 4E_u.$$

Note that $\Sigma(\partial_i u)^2 - (\partial_r u)^2 = |(R/r)u|^2$.

We give a detailed proof of this theorem as an exercise in the technique just described in Sections 6.1–6.4, following the same three Steps as before.

Proof:

Step 1. Establishing a differential identity. We write $2(\Box u)(\partial_r u)$ as a sum of terms in divergence form plus a sum Q of quadratic terms in $\partial_t u, \partial_i u$, just as in Sections 6.2 or 6.4. More precisely,

$$(\partial_t^2 u)(\partial_r u) = (\partial_t^2 u)(\Sigma\omega_i\partial_i u)$$
$$= \partial_t[(\partial_t u)(\partial_r u)] - \frac{1}{2}\Sigma\partial_i[\omega_i(\partial_t u)^2] + \frac{1}{2}(\partial_t u)^2\Sigma\partial_i\omega_i,$$

$$(\partial_j^2 u)(\partial_r u) = \partial_j[(\partial_j u)(\partial_r u)] - \frac{1}{2}\Sigma\partial_i[\omega_i(\partial_j u)^2] - \Sigma_i(\partial_j u)(\partial_i u)(\partial_j\omega_i)$$
$$+ \frac{1}{2}(\partial_j u)^2\Sigma\partial_i\omega_i.$$

Since

$$\partial_j(\omega_i) = r^{-1}(\delta_{ij} - \omega_i\omega_j), \quad \Sigma\partial_i\omega_i = \frac{2}{r},$$
$$\Sigma(\partial_i u)(\partial_j u)(\partial_j\omega_i) = r^{-1}[\Sigma(\partial_i u)^2 - (\partial_r u)^2],$$

we obtain, gathering the terms,

$$2(\Box u)(\partial_r u) = \partial_t[2(\partial_t u)(\partial_r u)] + \Sigma\partial_i\{\omega_i[\Sigma(\partial_j u)^2$$
$$- (\partial_t u)^2] - 2(\partial_i u)(\partial_r u)\} + Q,$$
$$Q = \frac{2}{r}[\Sigma(\partial_i u)^2 - (\partial_r u)^2] + \frac{2}{r}[(\partial_t u)^2 - \Sigma(\partial_i u)^2].$$

We note here a special feature of great importance: the sum Q of the quadratic terms in ∂u contains two different terms. The first term is nonnegative and gives a control of $(R/r)u$. The second term is an expression *with no special sign*, but which is of the form

$$a[(\partial_t u)^2 - \Sigma(\partial_i u)^2],$$

with $a = 2/r$. The following lemma allows us to further transform this expression.

Lemma 6.9. *The following identity holds:*

$$au(\Box u) = \partial_t[au\partial_t u - \frac{1}{2}(\partial_t a)u^2] - \Sigma\partial_i[au\partial_i u - \frac{1}{2}(\partial_i a)u^2] - a[(\partial_t u)^2$$
$$- \Sigma(\partial_i u)^2] + \frac{u^2}{2}\Box a.$$

We skip the straightforward proof, which uses the by now familiar technique of step 1. This lemma suggests considering the product

$$2(\Box u)\left(\partial_r u + \frac{u}{r}\right),$$

in which the bad quadratic terms cancel.

Step 2. Integration over the domain. Since we introduced the singular term u/r, we consider now the integral of $2(\Box u)(\partial_r u + \frac{u}{r})$ not in the full strip S_t, but in the domain

$$\{(x,s),\ 0 \le s \le t,\ |x| \ge \epsilon > 0\},$$

which is the exterior of a thin cylinder. The upper and lower boundary terms are the integrals

$$\int_{|x|\ge\epsilon} dx \left[\left(\partial_r u + \frac{u}{r}\right)(\partial_t u)\right]$$

taken at time t and 0, respectively. The lateral boundary term on the cylinder $|x| = \epsilon$ with outer normal $N = (-\omega, 0)$ is

$$\int \epsilon^{-1}\{\epsilon[(\partial_t u)^2 - \Sigma(\partial_i u)^2] + 2\epsilon(\partial_r u)^2 + 2u\partial_r u + \epsilon^{-1}u^2\}\epsilon^2 dt d\sigma_{S^2},$$

where $d\sigma_{S^2}$ is the area element on the unit sphere of $\mathbf{R}^3$. Letting ϵ go to zero, this last term goes to

$$4\pi \int_0^t u^2(0,s)ds.$$

Since $\Box u = 0$, there is no step 3 here. Since the multiplier $X = \partial_r$ is spacelike, the upper and lower boundary integrals have no special sign, and we have to control them. We write

$$2\left|\int (\partial_r u + \frac{u}{r})(\partial_t u)dx\right| \leq \int (\partial_t u)^2 + \int \left(\partial_r u + \frac{u}{r}\right)^2 dx,$$

and expand the last term. For the double product, we integrate by parts

$$2\int (\partial_r u)\frac{u}{r}dx = 2\int (\partial_r u)urdrd\omega = \int \partial_r (u^2)rdrd\omega = -\int \frac{u^2}{r^2}dx,$$

hence finally

$$2\left|\int (\partial_r u + \frac{u}{r})(\partial_t u)dx\right| \leq \int [(\partial_t u)^2 + (\partial_r u)^2]dx \leq 2E_u,$$

which completes the proof. $\Box$

6.6   KSS Inequality

The following inequality is named after its authors M. Keel, H. Smith and C. Sogge (See Notes).

Theorem 6.10. *There exists C such that, for all $u \in C^2(\mathbf{R}^3_x \times [0,T[)$ sufficiently decaying when $|x| \to +\infty$, and $t < T$,*

$$[\log(2+t)]^{-1/2}\left[\iint_{S_t} (1+r)^{-1}|\partial u|^2(x,s)dxds\right]^{1/2}$$

$$\leq CE_u^{1/2}(0) + C\int_0^t \|(\Box u)(\cdot,s)\|_{L^2}ds.$$

Proof: We only handle the case $\Box u = f$ with $u(x,0) = 0, (\partial_t u)(x,0) = 0$, the homogeneous case being handled similarly. In contrast with the preceding inequalities, the proof here does not involve a multiplier method and integrations by parts. It uses only the *standard* energy inequality, combined with three new ideas, which are **truncation, scaling, and dyadic decomposition.**

Step 1. The aim of this first step is to prove the inequality

$$\left[\int_0^t \int_{|x|\leq 2} |\partial u|^2 dxds\right]^{1/2} \leq C\int_0^t \|(\Box u)(\cdot,s)\|_{L^2}ds.$$

Note that this is a part of the inequality of Theorem 6.10 which is slightly better than the general inequality, since the factor $[\log(2 + t)]^{-1/2}$ is not there. The idea is to use a clever **truncation** process jointly with the **strong Huygens principle.** We define the disjoint regions R_k in the (x, t) space by

$$R_k = \{(x, t),\ k \leq |x| + t < k + 1,\ k = 0, 1, \ldots\}$$

and define f_k to be f in R_k and zero otherwise, thus $f = \Sigma f_k$. Let then v_k be the solution of the Cauchy problem

$$\Box v_k = f_k,\ v_k(x, 0) = 0,\ (\partial_t v_k)(x, 0) = 0.$$

Obviously, $u = \Sigma v_k$. The function v_k is zero for $t + |x| < k$ because the propagation speed is less than 1, and is also zero for $t \geq k + 1 + |x|$ by the strong Huygens principle. Hence, in the cylinder

$$|x| \leq 2,\ l \leq t \leq l + 1,$$

v_k vanishes identically unless $|k - l| \leq 2$. This implies, for some constant C, the inequality

$$\int_l^{l+1} \int_{|x| \leq 2} |\partial u|^2 dx dt = \int_l^{l+1} \int_{|x| \leq 2} |\Sigma_{|k-l| \leq 2} \partial v_k|^2 dx dt$$

$$\leq 5\Sigma_{|k-l| \leq 2} \int_l^{l+1} \int_{|x| \leq 2} |\partial v_k|^2 dx dt.$$

For each k, the standard energy inequality applied to v_k yields

$$\max_{t \leq l+1} ||\partial v_k(\cdot, t)||_{L^2}^2 \leq 4 \left(\int_0^{l+1} ||f_k(\cdot, t)||_{L^2} dt \right)^2.$$

Fixing $N \leq t$ and assuming $l \leq N - 1$, we thus obtain

$$\int_l^{l+1} \int_{|x| \leq 2} |\partial u|^2 dx dt \leq C\Sigma_{|k-l| \leq 2} \left(\int_0^N ||f_k(\cdot, t)||_{L^2} dt \right)^2.$$

Summing on l and taking the square roots, this yields

$$\left[\int_0^N \int_{|x| \leq 2} |\partial u|^2 dx dt \right]^{1/2} \leq C \left[\Sigma_{0 \leq k \leq N+1} \left(\int_0^N ||f_k(\cdot, t)||_{L^2} dt \right)^2 \right]^{1/2}.$$

In the righthand side of the above inequality, we recognize the standard euclidean norm in $\mathbf{R}^{N+2}$ of $\int_0^N A(t)dt$, where the vector $A(t)$ is

$$A(t) = (\|f_0(\cdot, t)\|_{L^2}, \dots, \|f_{N+1}(\cdot, t)\|_{L^2}).$$

Now

$$\|A(t)\| = (\Sigma\|f_k(\cdot, t)\|_{L^2}^2)^{1/2} \le \|f(\cdot, t)\|_{L^2},$$

since the f_k have disjoint supports in x for fixed t. The norm of the integral of A being less than the integral of the norm of A, we obtain finally

$$\left[\int_0^N \int_{|x|\le 2} |\partial u|^2 dx dt\right]^{1/2} \le C \int_0^N \|f(\cdot, t\|_{L^2} dt.$$

We remark now that the above inequality holds in fact for any t instead of N:

$$\left[\int_0^t \int_{|x|\le 2} |\partial u|^2 dx ds\right]^{1/2} \le C \int_0^t \|f(\cdot, s)\|_{L^2} ds.$$

To see this, let N be the integer part of t; splitting

$$\int_0^t \int_{|x|\le 2} |\partial u|^2 dx ds = \int_0^N + \int_N^t \le C \left(\int_0^t \|f(\cdot, s)\|_{L^2} ds\right)^2$$

$$+ \max_{N\le s\le t} \int_{|x|\le 2} |\partial u|^2(x, s) dx$$

gives the result.

Step 2. The aim of this second step is to take advantage of the **homogeneity properties** of the wave equation to pass from the inequality of step 1 (controlling the solution in the cylinder $|x| \le 2$) to an inequality in an annulus

$$\left[\int_0^t \int_{R\le|x|\le 2R} |x|^{-1} |\partial u|^2 dx ds\right]^{1/2} \le C \int_0^t \|(\Box u)(\cdot, s)\|_{L^2} ds.$$

Again, this is a part of the inequality of Theorem 6.10 with no logarithmic factor. Note that the constant C does not depend on R! The idea is to use a **scaling argument,** which is as follows: Let v be any C^2 function with $\Box v = g$ and zero traces on $\{t = 0\}$. Then, for any $R > 0$, the function $w_R(y, \tau) = v(Ry, R\tau)$ is the solution of $\Box w_R = g_R$ with traces $(0, 0)$ and

$$g_R(y, \tau) = R^2 g(Ry, R\tau).$$

We write, using the change of variables $x = Ry, s = R\tau$,

$$\int_0^t \int_{R \le |x| \le 2R} |x|^{-1} |\partial v|^2 (x, s) dx ds$$

$$= \int_0^{t/R} \int_{1 \le |y| \le 2} R^3 |y|^{-1} |\partial v|^2 (Ry, R\tau)|^2 dy d\tau$$

$$\le R \int_0^{t/R} \int_{|y| \le 2} |\partial w_R|^2 (y, \tau) dy d\tau.$$

We now use for the function w_R the inequality proved in step 1 and obtain

$$\int_0^{t/R} \int_{|y| \le 2} |\partial w_R|^2 dy d\tau \le C \left(\int_0^{t/R} ||g_R(\cdot, \tau)||_{L^2} d\tau \right)^2.$$

From the definition of g_R we get

$$||g_R(\cdot, \tau)||_{L^2}^2 = \int |g_R(y, \tau)|^2 dy = R \int g(x, R\tau)^2 dx,$$

$$\int_0^{t/R} ||g_R(\cdot, \tau)||_{L^2} d\tau = R^{-1/2} \int_0^t ||g(\cdot, s)||_{L^2} ds.$$

Hence, finally,

$$\int_0^t \int_{R \le |x| \le 2R} |x|^{-1} |\partial v|^2 (x, s) dx ds \le C \left(\int_0^t ||g(\cdot, s)||_{L^2} ds \right)^2.$$

Step 3. Dyadic decomposition. To finish the proof, we remark first that

$$\int_0^t \int_{|x| \ge t} (1 + |x|)^{-1} |\partial u|^2 dx ds \le C \log(2 + t) \max_{0 \le s \le t} E_u(s),$$

so we need only prove

$$\int_0^t \int_{|x| \le t} (1 + |x|)^{-1} |\partial u|^2 dx ds \le C \log(2 + t) \left(\int_0^t ||f(\cdot, s)||_{L^2} ds \right)^2.$$

Let N be the smallest integer such that $t \le 2^{N+1}$. We split the integral into a sum of integrals on *dyadic shells*

$$\int_0^t \int_{|x| \le t} (1 + |x|)^{-1} |\partial u|^2 dx ds \le \int_0^t \int_{|x| \le 2^{N+1}} (1 + |x|)^{-1} |\partial u|^2 dx ds$$

$$= \int_0^t \int_{|x| \leq 1} (1 + |x|)^{-1} |\partial u|^2 dx ds$$

$$+ \Sigma_{0 \leq k \leq N} \int_0^t \int_{2^k \leq |x| \leq 2^{k+1}} (1 + |x|)^{-1} |\partial u|^2 dx ds.$$

Using $N + 2$ times the inequality of step 2, we have the same bound for each term, and the result follows. $\qquad \square$

It is important to understand precisely what Theorem 6.10 says: We have, for any $\alpha > 0$,

$$\int_{S_t \cap \{|x| \geq \alpha t\}} (1 + r)^{-1} |\partial u|^2 dx ds \leq C \int_0^t \frac{ds}{1 + s} \int |\partial u|^2 dx$$

$$\leq C(\max_{0 \leq s \leq t} E_u(s)) \log(2 + t).$$

In other words, the standard energy inequality would imply Theorem 6.10 if the integral on S_t on the lefthand side were replaced by an integral on $S_t \cap \{|x| \geq \alpha t\}$ only. Thus, Theorem 6.10 provides an improved behavior of *all derivatives* of the solution far away *inside the light cone* (that is, in regions $|x| \leq \alpha t$, $\alpha < 1$), while Theorem 6.3 provides an improved behavior of *some special derivatives* of the solution *close to the boundary* of the light cone (that is, in regions $|x| \geq \alpha t$, $\alpha > 0$).

6.7   Conformal Inequality

We turn now to an energy inequality obtained, in the spirit of Section 6.4, using the nonspacelike multiplier

$$K_0 = (r^2 + t^2)\partial_t + 2rt\partial_r.$$

Theorem 6.11. *For $n \geq 3$ there exists C such that, for all $u \in C^2(\mathbf{R}_x^n \times [0, T[)$, sufficiently decaying when $|x| \to +\infty$, and all $t < T$,*

$$E_u^c(t)^{1/2} \leq CE_u^c(0)^{1/2} + C \int_0^t \|g(\cdot, s)\|_{L^2} ds.$$

Here, $g = (r^2 + t^2)^{1/2} \square u$, and the conformal energy E^c is defined by

$$E^c(t) = E_u^c(t) = \frac{1}{2} \int [(Su)^2 + |Hu|^2 + |Ru|^2 + u^2](x, t) dx.$$

Recall from Chapter 5 the definitions of the Lorentz fields

$$S = t\partial t + r\partial_r, \ H_i = t\partial_i + x_i\partial_t, \ R = x \wedge \partial.$$

Proof: The first step of the proof is as in Sections 6.1–6.4.

Step 1. Establishing a differential identity. We write $(\Box u)(K_0 u)$ as the sum Q of quadratic terms in ∂u and terms in divergence form:

$$(r^2 + t^2)(\partial_t^2 u)(\partial_t u) = \frac{1}{2}\partial_t[(r^2 + t^2)(\partial_t u)^2] - t(\partial_t u)^2,$$

$$(r^2 + t^2)(\partial_i^2 u)(\partial_t u) = \partial_i[(r^2 + t^2)(\partial_i u)(\partial_t u)] - \frac{1}{2}\partial_t[(r^2 + t^2)(\partial_i u)^2]$$
$$- 2x_i(\partial_t u)(\partial_i u) + t(\partial_i u)^2,$$

$$tx_i(\partial_t^2 u)(\partial_i u) = \partial_t[tx_i(\partial_t u)(\partial_i u)] - \frac{1}{2}\partial_i[tx_i(\partial_t u)^2]$$
$$- x_i(\partial_t u)(\partial_i u) + \frac{t}{2}(\partial_t u)^2,$$

$$tx_i(\partial_j^2 u)(\partial_i u) = \partial_j[tx_i(\partial_i u)(\partial_j u)] - \frac{1}{2}\partial_i[tx_i(\partial_j u)^2]$$
$$- t\delta_{ij}(\partial_i u)(\partial_j u) + \frac{t}{2}(\partial_j u)^2.$$

Gathering the terms, we obtain finally

$$(\Box u)(K_0 u) = \frac{1}{2}\partial_t[(r^2 + t^2)((\partial_t u)^2 + \Sigma(\partial_i u)^2) + 4rt(\partial_t u)(\partial_r u)]$$
$$+ \Sigma\partial_i[tx_i(-(\partial_t u)^2 + \Sigma(\partial_j u)^2) - (r^2 + t^2)(\partial_t u)(\partial_i u)$$
$$- 2rt(\partial_r u)(\partial_i u)] + (n - 1)t[(\partial_t u)^2 - \Sigma(\partial_i u)^2].$$

We see that Q reduces to a multiple of $(\partial_t u)^2 - \Sigma(\partial_i u)^2$. Just as in the proof of Morawetz inequality, we use Lemma 6.9 to tranform this term :

$$(n - 1)t[(\partial_t u)^2 - \Sigma(\partial_i u)^2] = -(n - 1)tu(\Box u) + \partial_t\left[(n - 1)tu(\partial_t u)\right.$$

$$\left.- \left(\frac{(n - 1)}{2}\right)u^2\right] - \Sigma\partial_i[(n - 1)tu\partial_i u].$$

Step 2. Integration over the domain. Integrating over the strip S_t, we obtain the identity

$$\int_{S_t} (\Box u)(K_0 u + (n - 1)tu)dxds = \tilde{E}_u(t) - \tilde{E}_u(0),$$

where the modified energy $\tilde{E}$ is

$$\tilde{E}_u(t) = \frac{1}{2}\int \{(r^2 + t^2)[(\partial_t u)^2 + \Sigma(\partial_i u)^2] + 4rt(\partial_t u)(\partial_r u)$$
$$+ 2(n-1)tu(\partial_t u) - (n-1)u^2\}dx.$$

This energy is the sum of two different contributions: The first two terms come from using the multiplier K_0 and are nonnegative, since K_0 is non-spacelike; this can be checked directly by writing

$$(r^2 + t^2)[(\partial_t u)^2 + (\partial_r u)^2 + |\left(\frac{R}{ra}\right)u|^2] + 4rt(\partial_t u)(\partial_r u)$$
$$= (Su)^2 + (t\partial_r u + r\partial_t u)^2 + (r^2 + t^2)|\left(\frac{R}{r}\right)u|^2.$$

The last two terms come from using Lemma 6.9 to get rid of the bad quadratic terms Q, and it is not immediately clear that they do not destroy the positivity of $\tilde{E}$. So we pause for a moment to discuss the sign of $\tilde{E}$.

Step 2'. Consider the simple identity $\Sigma\partial_i(x_i u^2) = nu^2 + 2ru\partial_r u$. Writing

$$2(n-1)ut\partial_t u = 2(n-1)u(Su - r\partial_r u)$$

and using the above identity, we obtain

$$\int [2(n-1)tu\partial_t u - (n-1)u^2]dx = \int [2(n-1)uSu + (n-1)^2 u^2]dx,$$

which gives

$$\tilde{E} = \frac{1}{2}\int \left\{(Su + (n-1)u)^2 + (t\partial_r u + r\partial_t u)^2 + (r^2 + t^2)\left|\left(\frac{R}{r}\right)u\right|^2\right\} dx.$$

Noticing that

$$|Hu|^2 = \Sigma(H_i u)^2 = (t\partial_r u + r\partial_t u)^2 + t^2[\Sigma(\partial_i u)^2 - \partial_r u)^2]$$
$$= (t\partial_r u + r\partial_t u)^2 + \frac{t^2}{r^2}|Ru|^2,$$

we finally obtain

$$\tilde{E} = \frac{1}{2}\int \{(Su + (n-1)u)^2 + |Hu|^2 + |Ru|^2\}dx.$$

This computation can be considered a true miracle! It proves that the modified energy is always nonnegative.

But we can obtain more. In fact, if we start writing the integrand of $\tilde{E}$ as a sum of squares, we get

$$(r^2 + t^2)[(\partial_t u)^2 + (\partial_r u)^2] + 4rt(\partial_t u)(\partial_r u) + 2(n-1)tu\partial_t u - (n-1)u^2$$
$$= (r^2 + t^2)[\partial_t u + \frac{t}{r^2 + t^2}(2r\partial_r u + (n-1)u)]^2 + (r^2 + t^2)(\partial_r u)^2$$
$$- (n-1)u^2 - \frac{t^2}{r^2 + t^2}(2r\partial_r u + (n-1)u)^2.$$

Setting $v = r^{(n-1)/2}u$ to take into account the volume element $r^{n-1}drd\omega$, we rewrite the last three terms above as

$$\frac{(r^2 - t^2)^2}{r^2 + t^2}(\partial_r v)^2 + (n-1)v^2(-1 + (n-1)\frac{r^2 + t^2}{4r^2}) - (n-1)\frac{r^2 + t^2}{r}v\partial_r v.$$

Assume now $n \geq 3$. Then $v = O(r)$ near $r = 0$, and we can integrate by parts the last term with respect to r between $r = 0$ and $r = +\infty$:

$$\int \frac{r^2 + t^2}{r}v\partial_r vdr = -\frac{1}{2}\int v^2\left(1 - \frac{t^2}{r^2}\right)dr.$$

If $n = 2$ and we assume that $v = O(r)$, the same computation is allowed. Suppose then that we take u radial, $\partial_t u$ appropriately chosen, and $v = O(r)$: since we already know that $\tilde{E} \geq 0$, the above computation gives us the **Poincaré inequality:**

$$\int_0^{+\infty} \frac{(r^2 - t^2)^2}{r^2 + t^2}(\partial_r v)^2 dr \geq \frac{1}{4}\int_0^{+\infty}\left(1 + \frac{t^2}{r^2}\right)v^2 dr.$$

Returning now to the case $n \geq 3$, we have automatically then that $v = O(r)$ near zero, and we can use the above Poincaré inequality to obtain

$$\tilde{E} \geq \frac{1}{2}\int\{(r^2 + t^2)|\left(\frac{R}{r}\right)u|^2 + \frac{(n-2)^2}{4}u^2\left(1 + \frac{t^2}{r^2}\right)\}dx.$$

Using again the miracle computation above, we obtain for some $C > 0$,

$$\tilde{E} \geq C\int[(Su)^2 + |Hu|^2 + |Ru|^2 + u^2]dx.$$

Step 3. Handling the remainder term. This last step is analogous to the usual one: We have to bound

$$\int_{S_t}(\Box u)(K_0 u + (n-1)tu)dxds.$$

To do this, we observe that

$$K_0 u = tSu + r(t\partial_r u + r\partial_t u),$$
$$(K_0 u + (n-1)tu)^2 \le 2(r^2 + t^2)[(Su + (n-1)u)^2 + (t\partial_r u + r\partial_t u)^2].$$

We apply the Cauchy–Schwarz inequality and get

$$\left| \int (\Box u)(K_0 u + (n-1)tu)dx \right| = \left| \int g[(r^2 + t^2)^{-1/2}(K_0 u + (n-1)tu)]dx \right|$$

$$\le C\|g\|_{L^2} \left(\int \{(Su + (n-1)u)^2 \right.$$

$$\left. + (t\partial_r u + r\partial_t u)^2\}dx \right)^{1/2},$$

where $g = (r^2 + t^2)^{1/2}\Box u$. Thanks to the miracle computation, we obtain

$$\left| \int (\Box u)(K_0 u + (n-1)tu)dx \right| \le C\|g\|_{L^2}\tilde{E}(t)^{1/2}.$$

The rest of the proof is then straightforward.   □

6.8   Exercises

1. Consider a function $u \in C^2(\mathbf{R}_x^n \times [0, T])$. Compute the energy $E_u(t)$ corresponding to the multiplier $Xu = \partial_t u + \alpha \partial_1 u$, in the spirit of Section 6.4; for which values of α is this energy positive definite? nonnegative? Answer the same questions for the multiplier $Xu = \partial_t u + \alpha \partial_r u$.

2. Let $u \in C^2(\mathbf{R}_x^n \times [0, T[)$ be a solution of $\Box u = 0$ with Cauchy data $u_0, u_1 \in C^\infty$, vanishing for $|x| \ge M$. Set

$$J_i(t) = \int_{\mathbf{R}^n} (\partial_i u)(x, t)(\partial_t u)(x, t)dx,$$

$$K(t) = \int_{\mathbf{R}^n} (\Sigma x_i \partial_i u)(x, t)(\partial_t u)(x, t)dx,$$

$$F_i(t) = \frac{1}{2} \int_{\mathbf{R}^n} x_i[(\partial_t u)^2 + \Sigma(\partial_i u)^2](x, t)dx,$$

$$F(t) = \frac{1}{2} \int_{\mathbf{R}^n} |x|^2[(\partial_t u)^2 + \Sigma(\partial_i u)^2](x, t)dx.$$

Show that all these quantities are well defined. Prove

$$\partial_t J_i = 0, \ \partial_t F_i = -J_i, \ \partial_t F = -2K,$$

and compute $\partial_t K$. What can be said about the sign of $\partial_t K$ for t large? (Hint: Use Exercise 9)

3. Prove that the standard energy inequality is "scale invariant" in the following sense: Assume given a real function $u \in C^2(\mathbf{R}_x^n \times [0,T[)$, and set

$$\Box u = f, u(x,0) = u_0(x), (\partial_t u)(x,0) = u_1(x).$$

Define $u_\lambda(x,t) = u(\lambda x, \lambda t)$ $(\lambda > 0)$. Then

$$\Box u_\lambda = f_\lambda, \ u_\lambda(x,0) = u_{0\lambda}(x), \ (\partial_t u_\lambda)(x,0) = u_{1\lambda}(x).$$

Compute f_λ, $u_{0\lambda}$, and $u_{1\lambda}$. Write down the standard energy inequality for u_λ and translate it into an inequality for u in terms of f, u_0, u_1. What do you obtain?

4. Let $u \in C^2(\mathbf{R}_x^n \times [0,T])$ be real, and set

$$\Box u = f, \ u(x,0) = u_0(x), \ (\partial_t u)(x,0) = u_1(x).$$

For $R > 0$, write down the energy inequality for u in the truncated light cone

$$C_R = \{(x,t), \ 0 \le t \le T, \ |x| \le R + T - t\}.$$

Deduce from this that if

$$\int \{|\nabla u_0|^2 + u_1^2\}(x)dx < \infty, \ \int_0^T \|f(\cdot,s)\|_{L^2(\mathbf{R}^n)}ds < \infty,$$

the function u satisfies the standard energy inequality in the full strip $\mathbf{R}_x^n \times [0,T]$.

5.(a) Let $u \in C^2(\mathbf{R}_x^n \times [0,T])$ be a solution of the equation

$$Pu \equiv \Box u + \alpha_0 \partial_t u + \Sigma \alpha_i \partial_i u = f,$$

where α_0, $\alpha_i \in C^0(\mathbf{R}_x^n \times [0,T])$. In order to obtain an inequality for u in terms of its traces u_0, u_1, and f, apply the standard inequality to

$$\Box u = f - \alpha_0 \partial_t u - \Sigma \alpha_i \partial_i u.$$

Setting
$$A(t) = \max_x |\alpha_0(x,t)| + \Sigma \max_x |\alpha_i(x,t)|, \ t \leq T,$$

use the Gronwall lemma to obtain the inequality

$$\max_{0 \leq s \leq t} E_u(s)^{1/2} \leq \left[E_u(0)^{1/2} + \sqrt{2} \int_0^t \|f(\cdot, s)\|_{L^2} ds \right] \exp\left(2 \int_0^t A(s) ds \right).$$

(b) A rougher way of handling the same problem is to prove a weighted inequality. To do this, compute as usual

$$\int_{D_T} (Pu)(\partial_t u) e^{-2\lambda t} dx dt,$$

where λ is a constant. Show that if $\lambda \geq \max_{0 \leq t \leq T} A(t)$, then

$$\sup_{0 \leq t \leq T} [e^{-\lambda t} E_u(t)^{1/2}] \leq E_u(0)^{1/2} + \sqrt{2} \int_0^T e^{-\lambda t} \|f(\cdot, t)\|_{L^2} dt.$$

(c) Suppose now that P contains a zero order term

$$Pu \equiv \Box u + \alpha_0 \partial_t u + \Sigma \alpha_i \partial_i u + \beta u.$$

What kind of inequality can be obtained for u?

6.(a) Consider the semilinear wave equation

$$\Box u = F(\partial_t u, \partial_1 u, \ldots, \partial_n u),$$

where $u \in C^2$ is assumed to be real and F is a C^1 real function of its arguments. Let

$$C_{(x_0, t_0)} = \{(x,t), 0 \leq t \leq t_0, |x - x_0| \leq t_0 - t\}$$

be a closed (truncated) light cone. Assume that $u, v \in C^2(C_{(x_0, t_0)})$ are two solutions of the equation, with the same traces

$$u(x,0) = v(x,0), \ (\partial_t u)(x,0) = (\partial_t v)(x,0).$$

Show that $u = v$ in $C_{(x_0, t_0)}$.

(b) Assume you are given two real functions $u_0, u_1 \in C^\infty(\mathbf{R}^n)$. Let us call a closed light cone $C_{(x_0, t_0)}$ "admissible" is there exists $u \in C^2(C_{(x_0, t_0)})$, solution of the equation with traces u_0, u_1 on $\{t = 0\}$. Consider Ω the union of all such admissible cones. Show that $\partial\Omega$ has slope equal to at most one and that there exists a unique solution u of the equation in Ω with traces u_0, u_1 on $\{t = 0\}$. The set Ω is called the maximal domain of determination of u_0, u_1 (the nonobvious fact is that Ω is not reduced to $\{t = 0\}$!).

7. Let $u \in C^2(\mathbf{R}_x^3 \times [0, \infty[)$ be a solution of the wave equation with traces $u_0, u_1 \in C_0^\infty(\mathbf{R}^3)$. Use the representation formula from Chapter 5 to prove

$$\left[\int (|\partial_t u|^p + \Sigma|\partial_i u|^p)(x, t)dx \right]^{1/p} \sim t^{(2/p)-1}, \ t \to +\infty.$$

Deduce from this that no "energy conservation" in the sense of L^p can hold for $p \neq 2$.

8. Let $u \in C^3(\mathbf{R}_x^n \times [0, \infty[)$ $(n \geq 3)$, and set

$$\Box u = f, \ u(x, 0) = u_0(x), \ (\partial_t u)(x, 0) = u_1(x).$$

Assume that u_0 and u_1 are C^∞ functions vanishing for $|x| \geq M$. Suppose we want to obtain a control of $||(\partial Zu)(\cdot, t)||_{L^2}$, for all the Lorentz fields Z. We can either

(a) commute the derivatives ∂_t, ∂_i with the equation $\Box u = f$, and then apply the conformal energy inequality, or

(b) commute the Lorentz fields Z with the equation $\Box u = f$, and then apply the standard energy inequality.

Compare the two inequalities thus obtained.

9. Let $u \in C^2(\mathbf{R}_x^3 \times [0, \infty[)$ satisfy a perturbed wave equation

$$\Box u + \alpha(x, t)Zu = 0, \ u(x, 0) = u_0(x), \ (\partial_t u)(x, 0) = u_1(x),$$

where $\alpha \in C^0$ is given and Z is a Lorentz field (say $Z = S, R_i$, or H_i). One wants to obtain an energy inequality for u; assume, for simplicity, that both traces u_0, u_1 have compact supports.

(a) A first possibility is to use the conformal energy inequality, which, precisely, gives a control of the Lorentz fields. Show that if $t \max_x |\alpha(x, t)| \in L_t^1$, then, for some C and all t, $\Sigma||(Zu)(\cdot, t)||_{L^2} \leq C$.

(b) A more subtle strategy is to use the improved standard energy inequality, which gives an additional control of the special derivatives $T_i = \partial_i + \omega_i \partial_t$.

Prove the following identities, connecting the Lorentz fields with the fields T_i:

$$\partial_t + \partial_r = \Sigma \omega_i T_i, \ S = t\Sigma \omega_i T_i + (r - t)\partial_r,$$
$$R = x \wedge T, \ H_i = tT_i + \omega_i(r - t)\partial_t.$$

Prove the pointwise bound

$$|Zu| \leq C(1 + t)\Sigma|T_i u| + C|r - t||\partial u|.$$

(c) Write

$$|\alpha||T_i u| = (|\alpha| < r - t >^{1/2+\epsilon})(|T_i u| < r - t >^{-1/2-\epsilon})$$

and use the usual Cauchy–Schwarz argument to show that if

$$\|(1 + t)\max_x |\alpha| < r - t >^{1/2+\epsilon}\|_{L^2_t}$$

is small enough, the term

$$\int_0^T \|(1 + t)(\alpha T_i u)(\cdot, t)\|_{L^2} dt$$

can be absorbed in the lefthand side of the improved standard energy inequality. Show that, to handle the other term

$$\int_0^T \| < r - t > (\alpha \partial u)(\cdot, t)\|_{L^2} dt,$$

it is enough to have $\max_x | < r - t > \alpha| \in L^1_t$. Finally, show that this condition is implied by the previous condition on α. Compare with the condition obtained in (a).

10.   Let $u \in C^2(\mathbf{R}^3_x \times [0, \infty[)$ and assume (with the notation of Section 6.7)

$$E_u^c(0) + \int_0^{+\infty} \|g(\cdot, t)\|_{L^2} dt < \infty, \ g = (r^2 + t^2)^{1/2}\Box u.$$

Show that

$$\int [(\partial_t u)^2 - \Sigma(\partial_i u)^2]dx = O(t^{-1}), \ t \to +\infty.$$

(Hint: Write

$$(\partial_t u)^2 - \Sigma(\partial_i u)^2 = (\partial_t u - \partial_r u)(\partial_t u + \partial_r u) - |(R/r)u|^2,$$

and use the conformal energy identity, along with the identities from Chapter 5,

$$(r+t)(\partial_t + \partial_r) = S + \Sigma \omega_i H_i, \quad \frac{R}{r} = \left(\frac{1}{t}\right)(\omega \wedge H).$$

This phenomenon is called "equipartition of energy."

11.(a) Let $u \in C^\infty(\mathbf{R}_x^2 \times [0, \infty[)$ satisfy $\Box u = 0$. Give a simple sufficient condition on the traces

$$u_0(x) = u(x,0), \quad u_1(x) = (\partial_t u)(x,0)$$

of u, which implies $E_{Z^k u}(0) < \infty$ for all products $Z^k u$ of k Lorentz fields applied to u.

(b) Assume that the condition of (a) on the traces of u is satisfied. Show that, for $k \leq 2$,

$$\|(\partial Z^k u)(\cdot, t)\|_{L^2} < \infty,$$

and deduce from Klainerman inequality

$$(1 + r + t) < r - t >^{1/2} |\partial u|(x,t) \leq C.$$

In particular, in a region $r \leq Ct, C < 1$ inside the light cone,

$$|\partial u|(x,t) \leq \frac{C}{(1+t)^2}.$$

(This exercise shows that, by using energy inequalities, we do not obtain the optimal decay rate of the solution inside the light cone; compare with Exercise 10 of Chapter 5).

12. Let $u \in C^\infty(\mathbf{R}_x^n \times [0, \infty[)$, with

$$\Box u = f, \quad u(x,0) = 0, \quad (\partial_t u)(x,0) = 0, \quad f(x,0) = 0.$$

Assume that $f(\cdot, t)$ has zero spatial mean, in the sense that there exist n functions $g_i \in C^\infty$, vanishing for $|x| \geq M + t$ such that

$$f(x,t) = \Sigma \partial_i g_i(x,t), \quad g_i(x,0) = 0.$$

Define n functions v_i by

$$\Box v_i = g_i, \quad v_i(x,0) = 0, \quad (\partial_t v_i)(x,0) = 0.$$

Note that $u = \Sigma \partial_i v_i$.

(a) Show, for each i,

$$||(\partial_t^2 v_i)(\cdot, t)||_{L^2} + \Sigma||(\partial_j \partial_t v_i)(\cdot, t)||_{L^2} \leq C \int_0^t ||(\partial_t g_i)(\cdot, s)||_{L^2} ds.$$

(b) Using (a) and

$$g_i(x, t) = \int_0^t (\partial_t g_i)(x, s) ds,$$

show the inequalities

$$\begin{aligned}
||(\partial_j u)(\cdot, t)||_{L^2} &\leq \Sigma||(\partial_{ij}^2 v_i)(\cdot, t)||_{L^2} \\
&\leq C\Sigma||(\Delta v_i)(\cdot, t)||_{L^2} \\
&\leq C\Sigma(||(\partial_t^2 v_i)(\cdot, t)||_{L^2} + ||g_i(\cdot, t)||_{L^2}) \\
&\leq C\Sigma \int_0^t ||\partial_t g_i(\cdot, s)||_{L^2} ds.
\end{aligned}$$

(Hint: Use Plancherel formula to show

$$||\partial_{ij}^2 w||_{L^2} \leq ||\Delta w||_{L^2}.$$

(c) Deduce from (a) and (b)

$$||(\partial u)(\cdot, t)||_{L^2} \leq C\Sigma \int_0^t ||(\partial_t g_i)(\cdot, s)||_{L^2} ds.$$

Compare with a direct application of the standard energy inequality.
13.(a) In Section 6.4, we have associated to a general multiplier X and a domain D an energy density $e(N, X)$. Set

$$N = (N_1, \ldots, N_n, N_0), \quad \tilde{N} = (-N_1, \ldots, -N_n, N_0),$$

and define $\tilde{e}(\tilde{N}, X) = e(N, X)$. Prove the formula

$$\tilde{e}(\tilde{N}, X) = (Xu)(\tilde{N}u) - \frac{1}{2} < \tilde{N}, X > [(\partial_t u)^2 - \Sigma(\partial_i u)^2],$$

where the scalar product $<, >$ (defined in Exercise 5, Chapter 5), is

$$< (X_1, \ldots, X_n, X_0), (Y_1, \ldots, Y_n, Y_0) > = X_0 Y_0 - \Sigma X_i Y_i.$$

In particular, note that $\tilde{e}$ is *symmetric*.

(b) Prove (by a direct computation or geometric arguments) that if X and Y satisfy

$$< X, X > = 0, \quad < Y, Y > = 0, \quad X_0 > 0, \quad Y_0 > 0,$$

then $\tilde{e}(X, Y) \geq 0$.

(c) Consider two timelike vectors X, Y, with $X_0 > 0$, $Y_0 > 0$. In the plane spanned by X and Y, there are two vectors X' and Y' such that

$$< X', X' > \, = 0, \; < Y', Y' > \, = 0, \; X_0' > 0, \; Y_0' > 0.$$

Writing X and Y as linear combinations of X' and Y' with *positive* coefficients, conclude from (b) and (c) that $\tilde{e}(X, Y) \geq 0$. This proves Theorem 6.7.

14.(a) Let us consider the Klein–Gordon equation

$$\Box u + u = f.$$

Prove that the same energy inequality holds as for the wave equation, the standard energy being now

$$E_0(t) = \frac{1}{2} \int [(\partial_t u)^2 + \Sigma(\partial_i u)^2 + u^2] dx.$$

(b) Fix $M > 0$ and consider the Cauchy problem

$$\Box u + u = f, \; u(x, 2M) = u_0(x), \; (\partial_t u)(x, 2M) = u_1(x),$$

where u_0 and u_1 vanish for $\{|x| \geq M\}$, and f vanishes for $M + |x| \geq t$. Using the standard multiplier ∂_t, establish an energy inequality by computing

$$\int_{D_T} (\Box u + u)(\partial_t u) dx dt,$$

where, for $T \geq 2M$,

$$D_T = \{(x, t), \; t \geq 2M, \; t^2 - |x|^2 \leq T^2\}.$$

Show that the energy $E(T)$ obtained on the hyperboloid H_T, which forms the upper part of ∂D_T, can be written

$$E(T) = \frac{1}{2} \int_{H_T} [t^{-2}|Hu|^2 + \frac{t^2 - |x|^2}{t^2}(\partial_t u)^2 + u^2] dx.$$

Note that the vector fields H_i are tangent to the hyperboloid H_T. Prove the energy inequality

$$E(T)^{1/2} \leq E(2M)^{1/2} + \sqrt{2} \int_{2M}^{T} dt \left(\int_{H_t} f^2 dx \right)^{1/2}.$$

15.   Consider a function u as in Exercise 5.9(e), and its tranformed $v = u(I)$. Translate the standard energy inequality on v, using the multiplier ∂_T, into an energy inequality on u.

6.9 Notes

The standard energy inequality is proved in all textbooks. The improvement in Section 6.2 is to be found in Alinhac [2]. (See also Lindblad and Rodnianski [16]). Morawetz inequality is taken from [18]. We chose to include KSS inequality because of its simplicity and usefulness, and because its proof is typical of many such similar arguments in recent research. KSS inequality is taken from the original paper [13]. The presentation of the conformal energy inequality follows Hörmander [10].

Chapter 7

Variable Coefficient Wave Equations and Systems

7.1 What is a Wave Equation?

Consider, in $\mathbf{R}_x^n \times \mathbf{R}_t$, a second order partial differential operator of the form

$$L \equiv \partial_t^2 + 2\Sigma b_i(x,t)\partial_{it}^2 - \Sigma a_{ij}(x,t)\partial_{ij}^2 + L_1, a_{ij} = a_{ji},$$

where all coefficients are real and C^∞, and L_1 is a first order operator. We would like L to be an operator similar to the wave operator, and to enjoy the same properties: Finite speed of propagation, energy inequalities, etc. We saw in Chapter 2 that, for an operator in the plane, it is natural to require that its principal part should be the principal part of a product a real vector fields. Here, suppose first that L is homogeneous (that is, $L_1 \equiv 0$) with constant coefficients. For any $\xi \in \mathbf{R}^n$, we can construct an operator L_ξ by letting L act on functions of t and $s = x \cdot \xi$ only:

$$L(v(x \cdot \xi, t)) = [\partial_t^2 + 2(b \cdot \xi)\partial_{st}^2 - (\Sigma a_{ij}\xi_i\xi_j)\partial_s^2]v(s,t) = [L_\xi v](x \cdot \xi, t).$$

The requirement that L_ξ should be strictly hyperbolic as an operator in the plane (s,t) means

$$\delta = (b \cdot \xi)^2 + \Sigma a_{ij}\xi_i\xi_j > 0, \ \xi \neq 0.$$

This will motivate the following definitions.

S. Alinhac, *Hyperbolic Partial Differential Equations*, Universitext,
DOI 10.1007/978-0-387-87823-2_7, © Springer Science+Business Media, LLC 2009

Definition 7.1. *We call L hyperbolic (with respect to t) in a region* $\Omega \subset \mathbf{R}_x^n \times \mathbf{R}_t$ *if, for all* $(x,t) \in \Omega$ *and all* $\xi \in \mathbf{R}^n$,

$$\delta = (b(x,t) \cdot \xi)^2 + \Sigma a_{ij}(x,t)\xi_i\xi_j \geq 0.$$

In other words, L is hyperbolic if the roots $-\lambda_k(x,t,\xi)$ *(k = 1, 2) in* τ *of the characteristic equation*

$$\tau^2 + 2\Sigma b_i(x,t)\tau\xi_i - \Sigma a_{ij}(x,t)\xi_i\xi_j = 0$$

are real. The operator L is strictly hyperbolic if these roots are real and distinct for $\xi \neq 0$. *The* λ_k *are the "characteristic speeds" of the equation.*

Note that when no cross terms (that is, terms like $b_i\partial_{it}^2$) are present, the strict hyperbolicity of L is easily seen: this just means that the quadratic form $\Sigma a_{ij}\xi_i\xi_j$ is positive definite. In general, we can get rid (at least locally) of cross terms by the following procedure: let us write (with new first order terms L_1')

$$L = (\partial_t + \Sigma b_i\partial_i)^2 - \Sigma[a_{ij} + b_ib_j]\partial_{ij}^2 + L_1'.$$

If, in the region under consideration, we can perform a change of variables

$$X_1 = \phi_1(x,t), \ldots, X_n = \phi_n(x,t), \ T = t$$

in such a way that the vector field $\partial_t + \Sigma b_i\partial_i$ becomes ∂_T (see Chapter 1, Section 1.8), then the operator L takes the form

$$\bar{L} = \partial_T^2 - \Sigma\bar{a}_{ij}\partial_{X_iX_j}^2 + \bar{L}_1,$$

for some new coefficients $\bar{a}_{ij}$ and lower order terms $\bar{L}_1$.

From now on, we always tacitly assume that L is *strictly hyperbolic* in the region we consider.

7.2   Energy Inequality for the Wave Equation

We consider a domain D of the closed half-space $\mathbf{R}_x^n \times [0, \infty[$ exactly as in Chapter 6, Section 6.3. The following theorem gives an extension of the standard energy inequality in the case of a variable coefficients wave equation.

Theorem 7.2. *Let* $L = \partial_t^2 - \Sigma a_{ij}(x,t)\partial_{ij}^2$ *be a strictly hyperbolic wave equation. Assume that there exists* $0 < \alpha_0 \leq 1$ *such that, for all* $(x,t) \in D$, $\xi \in \mathbf{R}^n$,

$$\Sigma a_{ij}(x,t)\xi_i\xi_j \geq \alpha_0|\xi|^2.$$

Assume that the outward normal $N = (N_1, \ldots, N_n, N_0)$ on the upper part of ∂D satisfies

$$N_0^2 \geq \Sigma a_{ij} N_i N_j.$$

Then, for all $u \in C^2(\bar{D})$ sufficiently decaying as $|x| \to +\infty$,

$$\max_{0 \leq s \leq t} E_u(s)^{1/2} \leq [E_u(0)^{1/2} + \sqrt{2} \int_0^t \|f(\cdot, s)\|_{L^2} ds] \exp \left(\frac{2}{\alpha_0} \int_0^t A(s) ds \right).$$

Here, $f = Lu$, the energy $E_u(t)$ is defined as

$$E_u(t) = \frac{1}{2} \int_{\Sigma_t} [(\partial_t u)^2 + \Sigma a_{ij}(\partial_i u)(\partial_j u)](x, t) dx,$$

and the amplification factor A is

$$A(t) = \max_x [\Sigma |\partial_t a_{ij}| + \Sigma |\partial_k a_{ij}|](x, t).$$

Proof: The steps of the proof are exactly the same as for the constant coefficients case. Once again, we restrict ourselves to the case of a real function u.

Step 1. Establishing a differential identity. We write the product $(Lu)(\partial_t u)$ as usual:

$$(\partial_t^2 u)(\partial_t u) = \frac{1}{2} \partial_t [(\partial_t u)^2],$$
$$(a_{ij} \partial_{ij}^2 u)(\partial_t u) = \partial_i [a_{ij}(\partial_j u)(\partial_t u)] - a_{ij}(\partial_j u)(\partial_{ti}^2 u) - (\partial_i a_{ij})(\partial_t u)(\partial_j u).$$

To handle the second term above, we use the symmetry of $a_{ij} = a_{ji}$:

$$a_{ij}(\partial_j u)(\partial_{ti}^2 u) = \partial_t [a_{ij}(\partial_i u)(\partial_j u)] - (\partial_t a_{ij})(\partial_i u)(\partial_j u) - a_{ij}(\partial_i u)(\partial_{tj}^2 u).$$

Summing over i, j gives us

$$2\Sigma a_{ij}(\partial_j u)(\partial_{ti}^2 u) = \partial_t [\Sigma a_{ij}(\partial_i u)(\partial_j u)] - \Sigma(\partial_t a_{ij})(\partial_i u)(\partial_j u).$$

Gathering the terms, we obtain

$$2(Lu)(\partial_t u) = \partial_t [(\partial_t u)^2 + \Sigma a_{ij}(\partial_i u)(\partial_j u)] + \Sigma \partial_i [-2\Sigma a_{ij}(\partial_j u)(\partial_t u)] + Q,$$
$$Q = -\Sigma(\partial_t a_{ij})(\partial_i u)(\partial_j u) + 2\Sigma(\partial_i a_{ij})(\partial_j u)(\partial_t u).$$

Step 2. Integration on the domain. We integrate in the truncated domain D_t using Stokes formula:

$$\int_{D_t} (Lu)(\partial_t u) dx ds = E_u(t) - E_u(0) + \int_{\Lambda_t} I d\sigma + \int_{D_t} Q dx ds.$$

The boundary integrand I is

$$2I = N_0[(\partial_t u)^2 + \Sigma a_{ij}(\partial_i u)(\partial_j u)] - 2(\partial_t u)\Sigma a_{ij} N_i \partial_j u.$$

Introducing the vector fields

$$\bar{\partial}_i = \partial_i - \left(\frac{N_0}{N_i}\right)\partial_t,$$

which span the tangent plane to ∂D, we can write

$$I = N_0 \Sigma a_{ij}\bar{\partial}_i u \bar{\partial}_j u + N_0^{-1}(\partial_t u)^2 (N_0^2 - \Sigma a_{ij} N_i N_j).$$

The assumptions of the theorem imply $I \geq 0$.

Step 3. Handling the remainder terms. We handle the term $\int (Lu)(\partial_t u)dxds$ exactly as before:

$$\left| \int (Lu)(\partial_t u)dxds \right| \leq \sqrt{2} \int_0^t \|f(\cdot,s)\|_{L^2} E_u(s)^{1/2} ds.$$

It remains to deal with the "error term" $\int_{D_t} Q dxds$. We have

$$|Q| \leq 2A[(\partial_t u)^2 + \Sigma(\partial_i u)^2] \leq 2\left(\frac{A}{\alpha_0}\right)[(\partial_t u)^2 + \Sigma a_{ij}(\partial_i u)(\partial_j u)],$$

hence by integration

$$\left| \int_{D_t} Q dxds \right| \leq \frac{2}{\alpha_0} \int_0^t A(s)E_u(s)ds.$$

We want to apply the Gronwall lemma (see Chapter 2, Lemma 2.16) to the inequality we have obtained so far:

$$E(t) \leq E(0) + \sqrt{2} \int_0^t \|f(\cdot,s)\|_{L^2} E(s)^{1/2} ds + \frac{2}{\alpha_0} \int_0^t A(s)E(s)ds.$$

We use the same trick as before: For $t' \leq t < T$, we write

$$E(t') \leq E(0) + \sqrt{2} \int_0^t \|f(\cdot,s)\|_{L^2} E(s)^{1/2} ds + \frac{2}{\alpha_0} \int_0^{t'} A(s)E(s)ds.$$

Applying now the lemma, we get

$$\max_{0 \leq t' \leq t} E(t') \leq [E(0) + \sqrt{2} \int_0^t \|f(\cdot,s)\|_{L^2} E(s)^{1/2} ds] \exp\left(\frac{2}{\alpha_0} \int_0^t A(s)ds\right).$$

This implies immediately the conclusion.   □

7.3   Symmetric Systems

7.3.1   Definitions and Examples

Just as in Chapter 2, it is useful to consider also first order $N \times N$ systems

$$L = S(x,t)\partial_t + \Sigma A_i(x,t)\partial_i + B(x,t).$$

Here, S, A_i, and B are C^∞ $N \times N$ matrices, and the operator acts on functions

$$u : \mathbf{R}_x^n \times \mathbf{R}_t \supset \Omega \to \mathbf{C}^N.$$

We will always assume S to be invertible. To define hyperbolicity, we can follow the same path as in Section 1: Assuming $B = 0$ and constant coefficients, we let L act on functions of t and $s = x \cdot \xi$ only; we thus define a system L_ξ

$$L_\xi v(s,t) = S\partial_t v + (\Sigma A_i \xi_i)\partial_s v.$$

For this system to be hyperbolic in the (s,t)-plane, we require $S^{-1}\Sigma A_i \xi_i$ to have only real eigenvalues. This motivates the following definitions.

Definition 7.3. *We call L hyperbolic (with respect to t) in the region $\Omega \subset \mathbf{R}_x^n \times \mathbf{R}_t$ if, for all $(x,t) \in \Omega$ and $\xi \in \mathbf{R}^n$, the matrix $S^{-1}\Sigma A_i \xi_i$ has real eigenvalues. The operator L is strictly hyperbolic if these eigenvalues are also distinct for $\xi \neq 0$.*

Definition 7.4. *The system L is symmetric if the matrices S and A_i are hermitian. The operator is symmetric hyperbolic if, moreover, S is positive definite.*

We will see below why symmetry is an *essential feature* to obtain energy inequalities. Let us comment about the concept of symmetric hyperbolic system: If $S = id$, symmetry clearly implies hyperbolicity, since an hermitian matrix has real eigenvalues. More generally, let $X \neq 0$ such that $S^{-1}AX = \lambda X$: This implies

$$^t\bar{X}AX = \lambda\,^t\bar{X}SX.$$

Since $^t\bar{X}SX > 0$, λ is the quotient of two real quantities, hence λ is real. In other words, a symmetric hyperbolic operator is actually hyperbolic. Note that an operator may be symmetric hyperbolic without being strictly hyperbolic, a fact which is important in applications.

Example 7.5. Just as in Chapter 2, we can reduce a variable coefficients wave equation

$$P = \partial_t^2 - \Sigma a_{ij}\partial_{ij}^2$$

to a first order symmetric system. To do this, set $u_0 = \partial_t u$, $u_i = \partial_i u$; then the $n + 1$ equations

$$\partial_t u_0 = \Sigma a_{ij} \partial_j u_i, \ \Sigma_j a_{ij} \partial_t u_j = \Sigma_j a_{ij} \partial_j u_0$$

form a symmetric system L for which

$$S_{00} = 1, \ S_{0i} = 0, \ S_{ij} = a_{ij}.$$

Note that the condition $a >> 0$ (this notation meaning "positive definite") of strict hyperbolicity of P is equivalent to the condition $S >> 0$ which makes L symmetric hyperbolic.

Example 7.6. The nonlinear system of compressible isentropic Euler equations is

$$\partial_t \rho + div(\rho u) = 0, \ \partial_t u_i + (\Sigma u_j \partial_j) u_i + \rho^{-1} \partial_i P = 0, \ i = 1, \dots, n.$$

Here, $\rho > 0$ is the density of the fluid, $u(x, t) \in \mathbf{R}^n$ is its velocity as observed at time t at the point x, and P is the pressure, considered here as a given function of ρ (depending on the particular physical fluid we are considering), of the form, say,

$$P = P(\rho) = A\rho^\gamma, \ A > 0, \ \gamma > 1.$$

If we introduce the new unknown function c instead of ρ,

$$c = \alpha \rho^{(\gamma-1)/2}, \ \frac{\alpha(\gamma - 1)}{2} = (A\gamma)^{1/2},$$

we obtain a new system

$$D_t c + \frac{c(\gamma - 1)}{2} \ div \, u = 0,$$

$$D_t u_i + \frac{c(\gamma - 1)}{2} \partial_i c = 0, \ i = 1, \dots, n,$$

where $D_t \equiv \partial_t + \Sigma u_i \partial_i$. For each $\bar{c}$ given, the linearized system

$$D_t c + \frac{\bar{c}(\gamma - 1)}{2} \ div \, u = 0, \ D_t u_i + \frac{\bar{c}(\gamma - 1)}{2} \partial_i c = 0, \ i = 1, \dots, n,$$

is symmetric hyperbolic, but not strictly hyperbolic as soon as $n \geq 3$.

Example 7.7. The Maxwell system is

$$\partial_t E + curl B = 0, \ \partial_t B - curl E = 0,$$

where E and B are, respectively, the electric field and the magnetic field. Considered as a 6×6 system with unknown $u = (E, B)$, it is a symmetric hyperbolic system.

7.3.2   Energy Inequality

Let us consider a domain $D \subset \mathbf{R}_x^n \times [0, \infty[$ as in Section 7.2, with outer normal $N = (N_1, \ldots, N_n, N_0)$. We have the following energy inequality.

Theorem 7.8. *Let*
$$L = S\partial_t + \Sigma A_i \partial_i + B$$

be a symmetric hyperbolic system in $\bar{D}$, and assume, for some constant $\alpha_0 > 0$ and all $(x, t) \in D$, $X \in \mathbf{C}^N$,

$$^t\bar{X}S(x,t)X \geq \alpha_0|X|^2.$$

Define the energy $E_u(t)$ by

$$E_u(t) = \frac{1}{2}\int_{\Sigma_t} [^t uSu](x,t)dx.$$

Assume that, on the upper part of ∂D, the hermitian matrix $N_0 S + \Sigma N_i A_i$ is nonnegative. Then, for all $u \in C^1(\bar{D})$ sufficiently decaying when $|x| \to +\infty$,

$$\max_{0 \leq s \leq t} E_u(s)^{1/2} \leq \left[E_u(0)^{1/2} + \left(\frac{2}{\alpha_0}\right)^{1/2} \int_0^t \|f(\cdot, s)\|_{L^2} ds \right] \exp\left(\frac{2}{\alpha_0}\int_0^t A(s)ds\right).$$

Here, the amplification factor A is

$$A(t) = \max_x \|C(x,t)\|, \; C = \frac{B + {}^t\bar{B}}{2} - \frac{1}{2}(\partial_t S + \Sigma \partial_i A_i).$$

Proof: The steps of the proof are again essentially the same as before. For simplicity, we assume that S, A_i, B, and u are real.

Step 1. Establishing a differential identity. We write $^t u(Lu)$ as a sum of terms in divergence form and quadratic terms in u; Using the symmetry of S and A_i we obtain

$$^t uS\partial_t u = \frac{1}{2}\partial_t[^t uSu] - \frac{1}{2}{}^t u(\partial_t S)u,$$

$$^t uA_i\partial_i u = \frac{1}{2}\partial_i[^t uA_i u] - \frac{1}{2}{}^t u(\partial_i A_i)u.$$

Gathering the terms,

$$2^t uLu = \partial_t[^t uSu] + \Sigma\partial_i[^t uA_i u] + 2^t uCu, \; C = \frac{B + {}^t B}{2} - \frac{1}{2}(\partial_t S + \Sigma\partial_i A_i).$$

Step 2. Integration over the domain. Integrating $^t u L u$ in the domain D_t, we obtain using Stokes formula

$$\int_{D_t} {}^t u L u \, dx ds = E_u(t) - E_u(0) + \int_{\Lambda_t} I d\sigma + \int_{D_t} {}^t u C u \, dx ds.$$

The lateral boundary terms integrand I is given by

$$2I = {}^t u (N_0 S + \Sigma N_i A_i) u.$$

The assumption of the theorem implies $I \geq 0$.

Step 3. Handling the remainder term. We handle now the two remaining integrals on D_t. Since

$$|{}^t u C u| \leq ||C|| \, |u|^2 \leq \frac{||C||}{\alpha_0} ({}^t u S u),$$

$$\left| \int_{D_t} {}^t u C u \, dx ds \right| \leq \frac{2}{\alpha_0} \int_0^t A(s) E(s) ds.$$

Also, as usual,

$$\left| \int_{D_t} {}^t u L u \, dx ds \right| \leq \left(\frac{2}{\alpha_0} \right)^{1/2} \int_0^t ||f(\cdot, s)||_{L^2} E(s)^{1/2} ds.$$

We have obtained so far the inequality

$$E(t) \leq E(0) + \left(\frac{2}{\alpha_0} \right)^{1/2} \int_0^t ||f(\cdot, s)||_{L^2} E(s)^{1/2} ds + \frac{2}{\alpha_0} \int_0^t A(s) E(s) ds.$$

The end of the proof is exactly similar to that for Theorem 7.2.   □

7.4   Finite Speed of Propagation

The two theorems of Sections 7.2, 7.3 have important corollaries, which display the finite speed of propagation property for hyperbolic wave equations or symmetric hyperbolic systems. We give here the simplest form of this corollary for the wave equation.

Theorem 7.9. *Let us consider in* $\mathbf{R}_x^n \times [0, \infty[$ *a hyperbolic wave equation*

$$Lu = \partial_t^2 u - \Sigma a_{ij} \partial_{ij}^2 u + a_0 \partial_t u + \Sigma a_i \partial_i u + bu.$$

Assume that the coefficients a_{ij} are C^1, and that, for some constant $\lambda_0 > 0$ and all $\xi \in \mathbf{R}^n$, $\xi \neq 0$, they satisfy everywhere

$$0 < \Sigma a_{ij}(x,t)\xi_i\xi_j \leq \lambda_0^2|\xi|^2.$$

Assume that the other coefficients a_0, a_i, b are bounded. Consider $u \in C^2$ $(\mathbf{R}_x^n \times [0,\infty[)$ with

$$Lu = f, \; u(x,0) = u_0(x), \; (\partial_t u)(x,0) = u_1(x).$$

Assume that u_0, u_1 vanish for $|x| \geq M$, and that f vanishes for $|x| \geq M + \lambda_0 t$. Then u vanishes for $|x| \geq M + \lambda_0 t$.

Proof: The proof is a typical application of the energy inequality of Theorem 7.2: In this theorem, we assume L to be homogeneous of order two. Here, we assume that lower order terms are present. Let $R > M$ and consider the (rotationally invariant) domain D_R defined by

$$D_R = \{(x,t), t \geq 0, M + \lambda_0 t \leq r = |x| \leq 2R - M - \lambda_0 t\}.$$

An outward normal N to the upper boundary of D_R is given by

$$N_i = \pm \frac{x_i}{|x|}, \; N_0 = \lambda_0,$$

the sign depending on which point of ∂D we are considering. In all cases, N satisfies the condition

$$N_0^2 = \lambda_0^2 \Sigma N_i^2 \geq \Sigma a_{ij} N_i N_j$$

of Theorem 7.2. Moreover, since D_R is compact, the assumptions of Theorem 7.2 are satisfied with some $\alpha_0 > 0$ and some $A < \infty$. When applying Theorem 7.2 to the domain D_R, we write

$$\partial_t^2 u - \Sigma a_{ij}\partial_{ij}^2 u = f - a_0\partial_t u - \Sigma a_i\partial_i u - bu = g.$$

Hence we obtain, for some C, the inequality

$$\phi(t) \equiv \max_{0\leq s\leq t} E_u(s)^{1/2} \leq C \int_0^t \|g(\cdot,s)\|_{L^2} ds.$$

It is easy to handle the derivatives of u in g :

$$\|a_0\partial_t u + \Sigma a_i\partial_i u\|_{L^2} \leq CE_u^{1/2}.$$

The control of bu is more delicate, since the energy inequality controls *only* the derivatives of u. We simply write in D_R, with various constants C depending on R,

$$u(x,t) = \int_0^t (\partial_t u)(x,s)ds, |u(x,t)|^2 \leq C \int_0^t (\partial_t u)^2(x,s)ds,$$

$$\int_{\Sigma_t} u^2(x,t)dx \leq C \int_0^t ds \int_{\Sigma_t} (\partial_t u)^2(x,s)dx \leq C \int_0^t E(s)ds \leq C \max_{0 \leq s \leq t} E(s).$$

Finally, the inequality gives us, since $f = 0$ in D_R,

$$\phi(t) \leq C \int_0^t E(s)^{1/2} + C \int_0^t \phi(s)ds \leq C \int_0^t \phi(s)ds.$$

Applying the Gronwall lemma yields finally $u = 0$ in D_R. Since R is arbitrary, this completes the proof. $\square$

We leave it to the reader as Exercise 7 to establish an analogue of this theorem for systems.

7.5   Klainerman's Method

Klainerman's method is an "energy inequality method" which allows one, in global situations, to obtain pointwise estimates of solutions to variable coefficients wave equations. The denomination "energy method" is chosen to emphasize the fact that no attempt is made to represent the solution by some formula, in contrast with the parametrix methods (see section 7.7). Not only this method gives a good qualitative information on the behavior of the solutions, but it is an essential tool in studying nonlinear perturbations of the wave equation.

Consider a variable coefficients wave equation L in $\mathbf{R}_x^n \times \mathbf{R}_t$. To emphasize the fact that L is a perturbation of the wave equation, we write, with $x_0 = t$,

$$L = \partial_t^2 - \Delta_x + \Sigma_{0 \leq i,j \leq n} a_{ij}(x,t)\partial_{ij}^2,$$

where the coefficients a_{ij} are supposed to be "small" (see step 2 below). There are three steps in the method:

Step 1. Commuting Lorentz fields. Let k be an integer at most equal to $(n+2)/2$, and compute

$$Z^k Lu = 0 = [Z^k, L]u + L(Z^k u).$$

Here Z^k means a product of k Lorentz fields (either ∂_t, ∂_i or S, H_i, or R_i), and the bracket is the *commutator* C defined by

$$Cu = [Z^k, L]u = Z^k(Lu) - L(Z^k u).$$

To understand the structure of the operator C, we have to note a number of elementary facts:

i) If $P = \Sigma_{|\alpha| \le p} a_\alpha(y)\partial_y^\alpha$ and $Q = \Sigma_{|\alpha| \le q} b_\alpha(y)\partial_y^\alpha$ are differential operators (in some $y \in \mathbf{R}^n$ variable) of orders p and q, respectively, then $[P, Q] \equiv PQ - QP$ is a differential operator of order at most $p + q - 1$, since the terms of order exactly $p+q$ in PQ and in QP cancel, because they are the same. Here, Z^k is of order k, L is of order 2, hence C is of order at most $k + 1$.

ii) We can write symbolically a formula analogous to Leibniz formula,

$$Z^k(ab) = \Sigma_{p+q=k}(Z^p a)(Z^q b),$$

where the righthand side means a sum of products of p fields applied to a times q fields applied to b, these fields being taken among the original fields of the product Z^k.

iii) Denoting by ∂^l a product of l ordinary derivatives, we have

$$[Z, \partial^l] = [Z, \partial]\partial^{l-1} + \cdots + \partial^p[Z, \partial]\partial^{l-1-p} + \cdots + \partial^{l-1}[Z, \partial].$$

But, as we saw in Chapter 5, $[Z, \partial]$ is either zero or a constant times ∂. Hence, omitting irrelevant constants, we can write symbolically

$$[Z, \partial^l] = \Sigma\partial^l.$$

iv) By induction on k, we prove now

$$[Z^k, \partial^l] = \Sigma\partial^l Z^{k-1}.$$

In fact, if $Z^{k+1} = Z_0 Z^k$,

$$[Z^{k+1}, \partial^l] = Z_0 Z^k \partial^l - \partial^l Z_0 Z^k = Z_0[\Sigma\partial^l Z^{k-1}] + [Z_0, \partial^l]Z^k = \Sigma\partial^l Z^k.$$

v) Finally, putting together the above remarks, we obtain

$$[Z^k, a\partial^l] = \Sigma_{1 \le p, p+q \le k}(Z^p a)\partial^l Z^q + a\partial^l Z^{k-1}.$$

We have proved the following lemma.

Lemma 7.10. *The commutator C is a linear combination, with irrelevant numerical coefficients, of terms $(Z^p a)\partial Z^r u$, $1 \le r \le k$, $p+r \le k+1$, where a stands for one of the coefficients a_{ij}.*

Step 2. Using an energy inequality. We assume now

$$\Sigma_{Z,\, k \le (n+2)/2} E_{Z^k u}(0) < \infty,$$

a condition that can be checked on the Cauchy data of u using the equation $Lu = 0$. This condition is certainly true in particular when u_0, u_1 are smooth with compact support (see also Exercise 11 of Chapter 6). Applying then an energy inequality for L to the equation $L(Z^k u) = -Cu$, and summing the results for all products Z^k, we get a control of

$$\Sigma_{Z,\, k \le (n+2)/2} E_{Z^k u}(t)^{1/2}$$

by a righthand side which contains only the time integral of $\|Cu\|_{L^2}$. If we assume for all coefficients $a = a_{ij}$

$$\int_0^{+\infty} \Sigma_{Z,\, p \le (n+2)/2} \|Z^p a(\cdot, t)\|_{L^\infty} dt < \infty,$$

the Gronwall lemma will ensure that, for some C independent of t,

$$\Sigma_{Z,\, k \le (n+2)/2} E_{Z^k u}(t) \le C.$$

Step 3. Using Klainerman's inequality. In the situation of step 2, we have a control of $\Sigma \|(\partial Z^k u)(\cdot, t)\|_{L^2}$. But the above commutation formula $[Z^k, \partial^l] = \Sigma \partial^l Z^{k-1}$, applied for $l = 1$, gives

$$\Sigma |Z^k(\partial u)| \le \Sigma_{p \le k} |\partial Z^p u|.$$

Hence Klainerman's inequality (See Theorem 5.12) yields

$$|\partial u|(x, t) \le C(1 + r + t)^{-(n-1)/2} < r - t >^{-1/2},$$

which was our goal.

Of course, there are many variations of this argument: In particular cases the commutator C may contains only terms of a special structure; It is possible to use, instead of the standard energy inequality for L, some improved or conformal inequality, etc.

7.6   Existence of Smooth Solutions

Theorem 7.11. *Consider as in Section 7.3 the Cauchy problem for a symmetric hyperbolic system with C^∞ coefficients*

$$Lu = S\partial_t u + \Sigma A_i \partial_i u + Bu = f, \ u(x,0) = u_0(x).$$

Let D be a compact domain of determination with base Σ_0 satisfying the assumptions of Theorem 7.8, and assume $u_0 \in C^k(\Sigma_0)$ and $f \in C^k(D)$ be given, $k \geq 4 + [n/2]$. Then there exists a unique solution $u \in C^\sigma(D)$, $\sigma = k - [n/2] - 3$.

Proof:

Step 1. Smoothing operators. We first start with some preliminary considerations about smoothing operators. Fix $\phi \in C_0^\infty(\mathbf{R}_x^n)$ a nonnegative even function ϕ with $\int \phi dx = 1$, vanishing for $|x| \geq 1$; the family of functions $\phi_\epsilon(x) = \epsilon^{-n}\phi(x/\epsilon)$ is an "approximation of the identity," since

$$\int \phi_\epsilon(x)dx = \int \phi(y)dy = 1$$

and ϕ_ϵ vanishes for $|x| \geq \epsilon$. We define an operator C_ϵ by

$$C_\epsilon v = \phi_\epsilon * v, \ C_\epsilon v(x) = \int \phi_\epsilon(x - y)v(y)dy.$$

• The operator C_ϵ is *smoothing* in the sense that it takes a locally L^1 function into a C^∞ function.

• It is formally self-adjoint in L^2 since

$$(u, C_\epsilon v)_{L^2} \equiv \int u(x)\bar{C_\epsilon}v dx = \int u(x)\phi_\epsilon(x - y)\bar{v}(y)dx dy$$

$$= \int \bar{v}(y) \int \phi_\epsilon(y - x)u(x)dx = (C_\epsilon u, v)_{L^2}.$$

• Using the Cauchy–Schwarz inequality, we have

$$(C_\epsilon v(x))^2 \leq \int \phi_\epsilon(x - y)v^2(y)dy,$$

$$\|C_\epsilon v\|_{L^2}^2 = \int (C_\epsilon v(x))^2 dx \leq \int v(y)^2 dy = \|v\|_{L^2}^2.$$

Similarly, we have

$$\|C_\epsilon v\|_{L^\infty} \leq \int \phi_\epsilon(x - y)|v(y)|dy \leq \|v\|_{L^\infty}.$$

- If v is continuous, $C_\epsilon v$ converges to v uniformly locally.

Finally, the following commutation lemma holds.

Lemma 7.12 (Commutation Lemma) *There exists C such that, for $A, v \in C^1(\mathbf{R}_x^n)$, vanishing for $|x| \geq R$, all j and all $\epsilon > 0$,*

$$||[C_\epsilon, A]\partial_j v \equiv C_\epsilon A \partial_j v - A C_\epsilon \partial_j v||_{L^2} \leq C||\nabla A||_{L^\infty}||v||_{L^2}.$$

Note that this lemma "gains" one derivative, the derivative originally acting on v being transferred to A. The commutator can be explicitly written

$$\int \phi_\epsilon(x - y)[A(y) - A(x)]\partial_j v(y) dy.$$

Integrating by parts, this gives

$$-\int \phi_\epsilon(x-y)(\partial_j A)(y)v(y)dy + \int \epsilon^{-n-1}(\partial_j \phi)\left(\frac{x-y}{\epsilon}\right)[A(y) - A(x)]v(y)dy.$$

The estimate for the first integral is clear, so we concentrate on the second. Since $A(y) - A(x) = B(x, y) \cdot (x - y)$ with $|B(x, y)| \leq ||\nabla A||_{L^\infty}$, the second integral can be written

$$\Sigma \int \epsilon^{-n} B_k(x, y)[z_k(\partial_j \phi)(z)]\left(z = \frac{x-y}{\epsilon}\right)v(y)dy,$$

and the estimate follows. This proves the lemma.   □

Step 2. Continuation of the proof of Theorem 7.11. After these preliminaries, we start the actual proof of the theorem. The idea is to replace the original Cauchy problem by the new problem

$$S\partial_t v + \Sigma A_i \partial_i C_\epsilon v + Bv = f, \ v(x, 0) = u_0(x),$$

the operator C_ϵ being the one defined above. We will show that this problem has a solution u_ϵ, and that these solutions are bounded *independently of* ϵ in some C^s.

(a) Suppose $D \subset\subset B(0, R) \times [0, T]$: We choose extensions $\tilde{u}_0 \in C_0^k(B(0, R))$ of u_0 and $\tilde{f} \in C_0^k(B(0, R) \times [0, T])$ of f; we also fix $\psi \in C_0^\infty(B(0, R))$ being 1 in a neighborhood of D. We claim that there exists a solution $u_\epsilon \in C_0^k(\mathbf{R}_x^n \times [0, T])$ of the Cauchy problem

$$S\partial_t u_\epsilon + \psi(x)\Sigma A_i \partial_i C_\epsilon u_\epsilon + Bu_\epsilon = \tilde{f}, \ u_\epsilon(x, 0) = \tilde{u}_0(x).$$

In fact, the operator $\partial_i C_\epsilon$ is, for fixed ϵ, a *bounded operator* in C^k, since

$$\partial_i C_\epsilon v(x) = \int \epsilon^{-n-1}(\partial_i \phi)\left(\frac{x-y}{\epsilon}\right) v(y) dy.$$

Hence the above new Cauchy problem can be viewed as a Cauchy problem for an ODE in the t-variable for functions with values in $C^k(\mathbf{R}^n_x)$.

(b) We proceed now to evaluate the tangential derivatives $v_\alpha = \partial_x^\alpha u_\epsilon$ for $|\alpha| = l \leq k-1$. To do this, we apply ∂_x^α to the equation satisifed by u_ϵ, and obtain

$$L_\epsilon v_\alpha \equiv S\partial_t v_\alpha + \psi \Sigma A_i C_\epsilon \partial_i v_\alpha = F_\alpha,$$

where, for fixed t, $\|F_\alpha(\cdot, t)\|_{L^2} \leq C + C\Sigma_{|\beta| \leq l} \|\partial_x^\beta u_\epsilon(\cdot, t)\|_{L^2}$.

(c) The essential point is this: We can obtain for the system $S\partial_t + \Sigma \psi A_i C_\epsilon \partial_i$ governing v_α an energy inequality *with fixed constants independent of* ϵ. In fact, $(,)$ being the L^2 scalar product in x,

$$\begin{aligned}
(v_\alpha, \psi A_j C_\epsilon \partial_j v_\alpha) &= -(\partial_j [C_\epsilon \psi A_j v_\alpha], v_\alpha)\\
&= -(C_\epsilon \partial_j (\psi A_j) v_\alpha, v_\alpha) - ([C_\epsilon, \psi A_j] \partial_j v_\alpha, v_\alpha)\\
&\quad - (\psi A_j C_\epsilon \partial_j v_\alpha, v_\alpha).
\end{aligned}$$

Hence

$$2(v_\alpha, F_\alpha) = \partial_t \left[\int {}^t v_\alpha S v_\alpha dx \right] + R,$$

with $|R(t)| \leq C\|v_\alpha(\cdot, t)\|_{L^2}^2$, thanks to the commutation lemma. Proceeding as usual, we obtain, for a fixed constant C, the standard energy inequality for first order systems

$$\|v_\alpha(\cdot, t)\|_{L^2} \leq C\|v_\alpha(\cdot, 0)\|_{L^2} + C\int_0^t \|F_\alpha(\cdot, s)\|_{L^2} ds.$$

(d) Summing all energy inequalities for all $\alpha, |\alpha| \leq k-1$, and using the Gronwall Lemma, we obtain $\Sigma\|\partial_x^\alpha u_\epsilon\|_{L^2} \leq C$. Using Sobolev lemma (Lemma 5.11), this gives a control of the C^s ($s = k-2-[n/2]$) norm in x for fixed t of $u_\epsilon(x, t)$, and using the system on u_ϵ, we finally get a control of the C^s norm in all variables of u_ϵ. Using Ascoli's theorem, we obtain a subsequence of u_ϵ, which we denote by u'_ϵ, converging in C^{s-1} to a function u. Note that the assumption on k implies $s-1 \geq 1$. Writing

$$C_\epsilon u'_\epsilon - u = C_\epsilon(u'_\epsilon - u) + C_\epsilon u - u,$$

we see that $C_\epsilon u'_\epsilon$ converges to u at least in C^1, hence u is a solution of the Cauchy problem

$$S\partial_t u + \psi\Sigma A_i\partial_i u + Bu = \tilde{f}, \ u(x,0) = \tilde{u}_0(x).$$

Restricting u to D, we obtain a solution of the original Cauchy problem, and we already know that this solution is unique. $\square$

The "flaw" of this proof is this: For a symmetric hyperbolic system in $\mathbf{R}^n_x \times \mathbf{R}_t$, we have no direct control of the C^k norm of the solution by the C^k norms of the data. We are forced to use L^2 norms and energy inequalities to control the solution, and this implies, via the Sobolev lemma, some loss of C^k regularity. An optimal result can only be obtained in the framework of Sobolev spaces, using distribution theory, and this is the reason why we made no attempt to optimize the loss of derivatives in the present result.

7.7   Geometrical Optics

We finish this chapter by presenting a method for obtaining explicit approximate solutions of variable coefficients equations. This method has an extremely wide range of applications, far beyond hyperbolic equations (local solvability of general equations, counterexamples, etc.). In the first two paragraphs we sketch the method in the general setting of any operator P in $\mathbf{R}^n$. Only in Section 7.7.3 do we come back to specific use of it for the hyperbolic Cauchy problem.

7.7.1   An Algebraic Computation

Let $P(x,\partial_x) = \Sigma\alpha_{jk}(x)\partial^2_{jk} + \beta_j(x)\partial_j + \gamma(x)$ be a second order differential operator with C^∞ coefficients on $\mathbf{R}^n_x$. The following lemma is obtained by a straightforward computation.

Lemma 7.13. *Let $a, \phi \in C^\infty(\mathbf{R}^n)$ and $\tau \in \mathbf{C}$ be given. Then*

$$\exp(-i\tau\phi)P[a(x)\exp(i\tau\phi(x))] = -\tau^2 a(x)p_m(x,\nabla\phi(x))$$
$$+ i\tau[(\partial_\xi p_m)(x,\nabla\phi(x))\partial_x a(x) + a(x)q(x)] + P(a).$$

Here, $p_m(x,\xi) = \Sigma\alpha_{jk}(x)\xi_j\xi_k$ is the principal symbol of P, and $q = P\phi - \gamma\phi$.

A similar lemma for operators of order m is left as Exercise 11.

7.7.2 Formal and Actual Geometrical Optics

We use the above Lemma 7.13 as follows: We think of τ as a big parameter (say $|\tau| \to +\infty$), and try to choose a and ϕ so that $e^{-i\tau\phi}P(ae^{i\tau\phi})$ is as small as possible:

First, to cancel the biggest term (i.e. the coefficient of τ^2), we look, in the region of interest, for a function ϕ (the **"phase"**) satisfying the so-called **"eikonal equation"**

$$p_m(x, \nabla\phi(x)) = 0.$$

This is a fully nonlinear first order equation exactly of the type studied in Chapter 3. This equation may not have any solution in general, and if it has, ϕ is not necessarily chosen real (so that the exponential factor $e^{i\tau\phi}$ may be big even if $\tau \in \mathbf{R}$). If p_m is real, however, the procedure of Chapter 3 provides us with at least a local real solution ϕ.

Second, to cancel the second term (i.e. the coefficient of τ), we choose a (the **"amplitude"**) to be solution of the first order *linear* equation

$$L_1 a \equiv (\partial_\xi p_m)(x, \nabla\phi(x))\partial a + qa = 0.$$

This equation is called **"transport equation."** We remark that the principal part of L_1 is the projection on $\mathbf{R}_x^n$ of the Hamiltonian field of p_m taken on the graph of $\nabla\phi$: these are precisely the objects which appear when applying the method of Chapter 3 to construct ϕ. Again, the transport equation need not have any solution in general, but it has if p_m and ϕ are real, according to Chapter 1.

Third, we can improve the procedure by trying *formally*

$$a(x) = a_0(x) + \frac{a_1(x)}{\tau} + \frac{a_2(x)}{\tau^2} + \cdots + \frac{a_k(x)}{\tau^k} + \cdots.$$

Choosing successively (if possible)

$$L_1 a_0 = 0, \ iL_1 a_1 + Pa_0 = 0, \ iL_1 a_2 + Pa_1 = 0, \ldots$$

we obtain $e^{-i\tau\phi}P(ae^{i\tau\phi}) \sim 0$, this symbol meaning that all coefficients of the various powers of τ vanish.

This is the **formal** approach to geometrical optics.

To define an **actual** approximate solution $u = ae^{i\tau\phi}$ of $Pu = 0$, we have now two choices

i) For some integer N, we take simply

$$a(x) = a_0(x) + \cdots + \frac{a_N(x)}{\tau^N}.$$

Then, locally in x,

$$e^{-i\tau\phi}P(ae^{i\tau\phi}) = O(\tau^{-N}).$$

ii) We use the Borel lemma (see Exercise 14).

Lemma 7.14 (Borel Lemma) *Let $a_n \in \mathbf{C}$ be a sequence of numbers. Then there exists a function $f \in C^\infty(\mathbf{R})$ such that*

$$f(x) \sim \Sigma a_n x^n, \ x \to 0.$$

This last line ("asymptotically equivalent") means that for all integers N,

$$f(x) - \Sigma_{0 \le n \le N} a_n x^n = O(x^{N+1}), \ x \to 0.$$

In the present case, we use an extension of Borel lemma for the variable $1/\tau \to 0$, x being a parameter. We thus obtain a C^∞ function $a(x, 1/\tau)$ such that

$$e^{-i\tau\phi}P(ae^{i\tau\phi}) \sim 0.$$

7.7.3   Parametrics for the Cauchy Problem

Let us assume now that the operator P is a variable coefficients strictly hyperbolic second order equation in $\mathbf{R}_x^n \times \mathbf{R}_t$. We will apply the theory outlined above to solve, in a neighborhood of the origin, say, the homogeneous Cauchy problem

$$Pu = 0, \ u(x, 0) = u_0(x), \ (\partial_t u)(x, 0) = u_1(x).$$

Step 1. We choose two *real* "phase" functions $\phi_\epsilon(x, t, \omega)$ ($\epsilon = \pm$), depending on the parameter $\omega \in \mathbf{R}^n$, $|\omega| = 1$, solutions of the Cauchy problem

$$\partial_t\phi_\epsilon(x, t, \omega) + \lambda_\epsilon(x, t, \partial_x\phi_\epsilon(x, t, \omega)) = 0, \ \phi_\epsilon(x, 0, \omega) = x \cdot \omega.$$

Here, $\lambda_- < \lambda_+$ are the characteristic speeds of the equation.

Step 2. Taking $\tau \in \mathbf{R}$, we choose two "amplitudes" $a_\epsilon(x,t,\omega,1/\tau)$ satisfying all the transport equations and such that

$$a_\epsilon\left(x,0,\omega,\frac{1}{\tau}\right) = 1.$$

The possibility of such a choice is a consequence of the constructive proof of the Borel lemma.

We obtain in this way two families of approximate solutions of $Pu = 0$,

$$v_\epsilon(x,t,\tau,\omega) = a_\epsilon\left(x,t,\omega,\frac{1}{\tau}\right)\exp(i\tau\phi_\epsilon(x,t,\omega)).$$

Step 3. The "trick" is now to take

$$\xi \in \mathbf{R}^n, \ \tau = |\xi|, \ \omega = \frac{\xi}{|\xi|},$$

and sum the corresponding solutions by writing some integral with respect to ξ. We thus choose two functions $A_-(x)$ and $A_+(x)$ (say in $\mathcal{S}$ for simplicity), and set

$$u_a(x,t) = (2\pi)^{-n}\Sigma_\epsilon \int e^{i\tau\phi_\epsilon(x,t,\omega)}a_\epsilon(x,t,\omega,\frac{1}{\tau})\hat{A}_\epsilon(\xi)(1-\chi(|\xi|))d\xi.$$

Here, $\chi \in C_0^\infty(\mathbf{R})$ is a truncation function which is 1 in a neighborhood of the origin, used only to avoid any trouble in the integral for ξ close to the origin. The subscript a stands for "approximate." Taking into account the choices made for ϕ_ϵ and a_ϵ, we get

i) $Pu_a \in C^\infty$,

ii) $u_a(x,0) - (A_- + A_+)(x) \in C^\infty$,

iii) $(\partial_t u_a)(x,0) = (2\pi)^{-n}\Sigma \int e^{ix\cdot\xi}[-i\tau\lambda_\epsilon(x,0,\omega) + (\partial_t a_\epsilon)(x,0,\omega,\frac{1}{\tau})]$ $\hat{A}_\epsilon(\xi)(1-\chi(|\xi|))d\xi$.

(i) results from the fact that, after applying P to the integrals, we get a similar integral with a_ϵ replaced by a rapidly decaying function as $|\xi| \to +\infty$. (ii) is just the Fourier inversion formula, and (iii) is a straightforward computation.

Claim 7.15. *Given functions $u_0, u_1 \in C_0^\infty(\mathbf{R}^n)$, we can choose the functions A_- and A_+ in such a way that, for x close to zero,*

$$u_a(\cdot,0) - u_0 \in C^\infty, \ (\partial_t u_a)(\cdot,0) - u_1 \in C^\infty.$$

The proof of this claim would require some insight into the theory of pseudodifferential operators, which is far beyond the scope of this book (the interested reader may consult the book by S. Alinhac and P. Gérard [5]). However, we can grasp the essential point if we assume that the "symbols"

$$s_\epsilon = -i\tau\lambda_\epsilon(x,0,\omega) + (\partial_t a_\epsilon)\left(x,0,\omega,\frac{1}{\tau}\right)$$

appearing in the formula for $\partial_t u_a(x,0)$ do not depend on x. In this case, to prove the claim, we just have to choose $\hat{A}_-$ and $\hat{A}_+$ such that, for $|\xi|$ big enough,

$$\hat{A}_- + \hat{A}_+ = \hat{u}_0, \quad s_-\hat{A}_- + s_+\hat{A}_+ = \hat{u}_1.$$

This is possible, since by strict hyperbolicity $\lambda_- \neq \lambda_+$, which implies $s_- \neq s_+$ for $|\xi|$ big enough.   □

To summarize, we have obtained u_a such that

$$Pu_a \in C^\infty, \quad u_a(\cdot,0) - u_0 \in C^\infty, \quad (\partial_t u_a)(\cdot,0) - u_1 \in C^\infty.$$

We say that u_a is a solution **modulo** C^∞ of the Cauchy problem

$$Pu = 0, \quad u(x,0) = u_0(x), \quad (\partial_t u)(x,0) = u_1(x),$$

and the formula defining u_a is called a **"parametrix"** of the Cauchy problem. Since we know (from Section 7.6) how to solve the Cauchy problem for P with C^∞ data, we see that the exact solution u of the Cauchy problem is the sum of the approximate solution u_a and a C^∞ function. It is thus possible to solve (locally at least) the Cauchy problem for non-smooth data, and to read from the formula defining the explicit approximate solution u_a the singularities (modulo C^∞) of the true solution.

7.8   Exercises

1.(a) Consider, in $\mathbf{R}_x^2 \times \mathbf{R}_t$, the operator

$$L = \partial_t^2 + 2\partial_{t1}^2 + \frac{1}{2}(\partial_1^2 - \partial_2^2).$$

Show that L is strictly hyperbolic with respect to t. Compute the energy $E_u(T)$ obtained by integrating $(Lu)(\partial_t u)$ in a strip S_T, and observe that E is not positive.

(b) Performing the change of variables

$$X_1 = x_1 - t, \quad X_2 = x_2, \quad T = t, \quad u(x_1, x_2, t) = v(x_1 - t, x_2, t),$$

compute the operator $\bar{L}$ such that

$$(Lu)(x_1, x_2, t) = (\bar{L}v)(x_1 - t, x_2, t).$$

Deduce from this what could be an appropriate multiplier X for the original operator L in order to obtain an energy inequality with a positive energy.

2. Let $u \in C^2(\mathbf{R}_x^n \times \mathbf{R}_t)$ satisfy the equation with real C^1 coefficients

$$Lu \equiv \partial_t^2 u - \Delta_x u + \Sigma_{0 \le i,j \le n} a_{ij}(x,t) \partial_{ij}^2 u = f, \quad x_0 = t,$$

and assume that, for each t, $u(x,t)$ vanishes for large $|x|$. Assume that the coefficients a_{ij} satisfy $\Sigma|a_{ij}| \le 1/2$.

Prove the energy inequality

$$\|(\partial u)(\cdot, t)\|_{L^2} \le 2\{\|(\partial u)(\cdot, 0)\|_{L^2} + \int_0^t \|f(\cdot, s)\|_{L^2} ds\} \exp\left(4 \int_0^t A(s) ds\right),$$

where the amplification factor is

$$A(t) = \Sigma_{0 \le i,j,k \le n} \max_x |\partial_i a_{jk}(x,t)|.$$

3. In Example 7.5, compute for given ξ, the roots in τ of the characteristic equation

$$det(\tau S + \Sigma A_i \xi_i) = 0$$

of the obtained system. Compare with the roots of the characteristic equation

$$\tau^2 - \Sigma a_{ij} \xi_i \xi_j = 0$$

of P.

4. In Example 7.6, compute for given ξ, the roots in τ of the characteristic equation of the linearized Euler system.

5. Consider in $S_T = \mathbf{R}_x^n \times [0, T[$ a symmetric hyperbolic system with real C^1 coefficients

$$L \equiv S \partial_t + \Sigma A_i \partial_i + B.$$

Assume that there exist constants $\alpha_0 > 0$ and $\lambda_0 > 0$ such that, in the strip S_T,

$$S \ge \alpha_0, \quad \lambda_0 S \ge \frac{1}{2}(\partial_t S + \Sigma \partial_i A_i) - \frac{B + {}^t B}{2}.$$

Show, for all $\lambda \geq \lambda_0$, all $t < T$, and all $u \in C^1(S_T)$ sufficiently decaying when $|x| \to +\infty$, the energy inequality

$$\max_{0 \leq s \leq t} E_u(s)^{1/2} \leq E_u(0)^{1/2} + \left(\frac{2}{\alpha_0}\right)^{1/2} \int_0^t \|(Lu)(\cdot, s)\|_{L^2} e^{-\lambda s} ds,$$

where the energy $E_u(t)$ is defined by

$$E_u(t) = \frac{1}{2} e^{-2\lambda t} \int [{}^t u S u](x, t) dx.$$

6.   Consider in $S_T = \mathbf{R}_x^n \times [0, T[$ a symmetric hyperbolic system with singular real coefficients

$$L = S\partial_t + \Sigma A_i \partial_i + \frac{B}{T - t}.$$

Assume that $S, A_i, B \in C^1(\bar{S}_T)$, and that there exist constants M, $\alpha_0 > 0$ such that

$$S \geq \alpha_0, \quad \left\|\frac{{}^t B + B}{2}\right\| + \|\partial_t S + \Sigma \partial_i A_i\| \leq M.$$

Show that there are constants C, γ_0 such that, for all $\gamma \geq \gamma_0$, all $t \leq T$ and $u \in C^2(S_T)$ sufficiently decaying when $|x| \to +\infty$,

$$\int_{S_t} |u|^2 (T - s)^{\gamma - 1} dx ds \leq C \int |u|^2 (x, 0) dx + C \int_{S_t} |Lu|^2 (T - s)^{\gamma + 1} dx ds.$$

7. Write down, for symmetric hyperbolic systems, a "finite speed of propagation" theorem analogous to Theorem 7.9.

8.(a) To a function $v \in C^2(\mathbf{R}_x^n)$ we associate the function

$$M(x, t) = (t^{n-1} \sigma_{n-1})^{-1} \int_{S(x,t)} v(y) d\sigma(y) = \frac{1}{\sigma_{n-1}} \int_{S^{n-1}} v(x + \omega t) d\sigma(\omega),$$

which is its means over the sphere of radius t centered at x (Here, S^{n-1} is the unit sphere in $\mathbf{R}^n$ and σ_{n-1} its area). Prove $M(x, 0) = v(x)$, $(\partial_t M)(x, 0) = 0$, and

$$\partial_t^2 M - \Delta_x M + (n - 1) t^{-1} \partial_t M = 0.$$

(b) Given $\alpha \in \mathbf{R}$, consider more generally, in $\mathbf{R}_x^n \times [0, \infty[$ the singular wave equation

$$L_\alpha u \equiv \partial_t^2 u - \Delta_x u + \left(\frac{\alpha}{t}\right) \partial_t u = 0.$$

Let Σ_0 be a smooth compact domain in $\mathbf{R}_x^n$, $r \in C^1(\Sigma_0)$ a nonnegative function vanishing on $\partial\Sigma_0$, and D be the domain in $\mathbf{R}_x^n \times [0, \infty[$ with base Σ_0 defined by

$$D = \{(x, t), x \in \Sigma_0, 0 \le t \le r(x)\}.$$

Let $u \in C^2(D)$ be a real function with $u(x, 0) = f(x)$, $(\partial_t u)(x, 0) = 0$. Assuming $u(x, r(x)) = f(x)$, prove the identity

$$2 \int_D (L_\alpha u)(\partial_t u)\,dxdt = 2\alpha \int_D \frac{1}{t}(\partial_t u)^2 dxdt + \int_{\Sigma_0} (1 - |\nabla r|^2)[(\partial_t u)(x, r(x))]^2 dx.$$

(c) Deduce from (b) that if $\alpha > 0$, $|\nabla r| \le 1$ and $L_\alpha u = 0$, then f is harmonic.

(d) State the theorem about harmonic functions that follows from (a), (b), and (c). (See also for more details Alinhac [1]).

9. Consider in $\mathbf{R}_x^3 \times [0, \infty[$ the wave equation

$$Lu \equiv (1 + c(x, t))\partial_t^2 u - \Delta_x u = 0.$$

Here, c is a real C^∞ coefficient, and assume that for some constant C, $\eta > 1$,

$$|c| \le \frac{1}{2}, \quad \Sigma_{Z, 0 \le k \le 3}|Z^k c| \le C(1 + t)^{-\eta}.$$

As usual, Z^k means a product of k Lorentz fields.

(a) Let $u \in C^2(\mathbf{R}_x^3 \times [0, \infty[)$ be a real solution of $Lu = 0$ with Cauchy data

$$u_0(x) = u(x, 0), \quad u_1(x) = (\partial_t u)(x, 0)$$

vanishing for $|x| \ge M$. Prove that u vanishes for $|x| \ge M + \lambda_0 t$ for some $\lambda_0 > 0$ to be computed explicitly.

(b) Establish an energy inequality for L with constants independent of t (that is, without amplification factor).

(c) Prove by induction the formulas

$$[Z^k, \square] = \Sigma_{l \le k-1}\alpha Z^l \square, \quad [Z^k, \partial_t^2] = \Sigma_{l \le k-1}\alpha Z^l \partial^2.$$

Here, α stands for various constants, and ∂ stands for ∂_t or ∂_i.

(d) Using Klainerman's method, prove that if $u \in C^5$ is a solution of $Lu = 0$ with Cauchy data vanishing for $|x| \geq M$ as in (a), then, for some C and all $(x, t) \in \mathbf{R}_x^3 \times [0, \infty[$,

$$|\partial u(x, t)| \leq C(1 + r + t)^{-1} < r - t >^{-1/2} .$$

10.(a) Let $u \in C^2(\mathbf{R}_x^3 \times \mathbf{R}_t)$ be a real solution of $\Box u = 0$. Fix γ and $T > 0$ and set $u(x, t) = (T - t)^{-\gamma} w(x, t)$. Write down the equation for w. Perform, in the cone

$$C = \{(x, t), \ 0 \leq t \leq T, \ |x| \leq T - t\},$$

the change of variables

$$s = -\log(T - t), \ y = \frac{x}{T - t}, \ w(x, t) = v(y, s).$$

Write down the equation for v obtained in the cylinder

$$-\log T \leq s < +\infty, \ |y| \leq 1.$$

(Hint: One obtains

$$Lv = \partial_s^2 v + 2\Sigma y_i \partial_{si}^2 v + \Sigma y_i y_j \partial_{ij}^2 v - \Delta_y v + (2\gamma + 1)\partial_s v + 2(\gamma + 1)\Sigma y_i \partial_i v$$
$$+ \gamma(\gamma + 1)v = 0.$$

(b) Using the multiplier $(1 - |y|^2)^\mu \partial_s v$, establish an energy inequality for L by choosing carefully μ. What are the corresponding multiplier and energy for $\Box$?

11.(a) Let $\phi \in C^\infty(\mathbf{R}_x^n)$ and $\tau \in \mathbf{C}$ be a parameter. Prove by induction on $|\alpha|$ the formula

$$e^{-i\tau\phi}\partial_x^\alpha(e^{i\tau\phi}) = (i\tau)^{|\alpha|}(\nabla\phi)^\alpha + \frac{1}{2}(i\tau)^{|\alpha|-1}\Sigma(\partial_{jk}^2\xi^\alpha)(\nabla\phi)\partial_{jk}^2\phi$$
$$+ O(|\tau|^{|\alpha|-2}), \ |\tau| \to +\infty.$$

(b) Let $P(x, \partial_x) = \Sigma_{|\alpha| \leq m} a_\alpha(x)\partial_x^\alpha$ be a differential operator of order m in $\mathbf{R}_x^n$. For a given function $a \in C^\infty(\mathbf{R}_x^n)$, write down the two main terms (as $|\tau| \to +\infty$) of

$$e^{-i\tau\phi}P(ae^{i\tau\phi}).$$

What are the eikonal equation and the transport equation in this case?

12. Taking for P the standard wave equation $\Box$, compare the approximate solution u_a constructed in Section 7.7.3 with the solution obtained by a partial Fourier transformation in Chapter 5.

13. Consider a first order $(N \times N)$-system in $\mathbf{R}^n_x$ with C^∞ coefficients

$$L = \Sigma A_i(x)\partial_i + B(x).$$

Making all simplifying assumptions you may need, explain how one can construct approximate *vector* solutions

$$u_a(x) = a(x, 1/\tau)e^{i\tau\phi(x)}$$

of this system, following the pattern of Section 7.7 (define an eikonal equation, a transport equation, etc.).

14.(a) Prove the Borel lemma. Hint: Try

$$f(x) = \Sigma a_n x^n \chi(\lambda_n x)$$

for $\chi \in C_0^\infty(\mathbf{R})$ being 1 close to zero, and a sufficiently increasing sequence $\lambda_n \to +\infty$.

(b) Prove the version of Borel lemma "with parameters" actually used in the construction in Section 7.7.

15.(a) Consider the standard wave equation $\Box$ in $\mathbf{R}^n_x \times \mathbf{R}_t$. Check that $\phi(x, t) = x_2 + t$ is a phase function and compute the corresponding transport equation L_1.

(b) Starting with $a_0(x_1)$ vanishing close to $x_1 = 0$, show that one can choose all coefficients $a_k(x, t)$ of the amplitude a, and the amplitude a itself, with the same property.

This construction shows that the Cauchy problem for the wave equation with data on $\{x_1 = 0\}$ certainly cannot be well-posed, since the data can be zero and $\Box u$ very small without u being small.

7.9 Notes

All inequalities discussed in Chapter 6 for the wave equation have analogues for variable coefficient wave equations, but these analogues are harder to find in the literature. The analogue of the standard inequality is taken from Hörmander [10]; analogues of the improved standard inequality and of Morawetz type inequalities are established in Alinhac [3]. Conformal inequalities in the framework of variable coefficients can be found in Hörmander [10] or Klainerman [14]. Hyperbolic symmetric systems are discussed in John's book [12]. Existence of smooth solutions, generally handled by duality, is not easy to prove in an elementary context; again, the given result is taken from John (with a different proof, however). Finally, geometrical optics and its application to constructing parametrices is discussed in Taylor [23] and Vainberg [24].

Appendix

We gather here without complete proofs some basic facts about ordinary differential equations and submanifolds of $\mathbf{R}^n$, which we use in the text and may not be included in all standard differential calculus courses.

A.1   Ordinary Differential Equations

We refer here to the book of Hubbard and West [11].

A.1.1   Cauchy Problem

We consider the Cauchy problem for systems of ordinary differential equations (ODE)
$$x'(t) = F(x(t), t), \ x(t_0) = x_0.$$

Here, $\Omega \subset \mathbf{R}_x^n$ is open, $I \subset \mathbf{R}$ is an interval, $m_0 = (x_0, t_0) \in \Omega \times I$, and

$$F : \Omega \times I \to \mathbf{R}_x^n$$

is a C^1 function. A solution is a function $x : I \supset J \to \Omega$ of class C^1, defined on a subinterval $J \ni t_0$ of I, satisfying the equation and the initial condition.

The Cauchy–Lipschitz (local existence) theorem can be stated as follows: Let $m_0 = (x_0, t_0) \in \Omega \times I$, and assume the existence of $a > 0$, $b > 0$, such that $R = \bar{B}(x_0, b) \times [t_0 - a, t_0 + a] \subset \Omega \times I$. Define $M = \max_R |F|$, $\alpha = \min(a, b/M)$.

Theorem A.1 (Cauchy–Lipschitz local existence theorem). *There exists a unique solution $x \in C^1(J), J = [t_0 - \alpha, t_0 + \alpha]$ of the Cauchy problem.*

Proof: The first step in the proof is to write the system in integral form

$$x(t) = x_0 + \int_{t_0}^{t} F(x(s), s)ds.$$

We now look for $x \in C^0(J)$ satisfying this equation, since then $x \in C^1(J)$ and x is a solution of the Cauchy problem:

Step 1. Set $R' = J \times \bar{B}(x_0, b)$. Suppose $y \in C^0(J)$ has its graph in R', and define for $t \in J$

$$x(t) = x_0 + \int_{t_0}^{t} F(y(s), s)ds.$$

Then also x has its graph in R', since $||x(t) - x_0|| \leq M|t - t_0| \leq b$.

Step 2. Using step 1, define a sequence of functions $x^n \in C^0(J)$ by

$$x^0 = x_0, \ x^{n+1}(t) = x_0 + \int_{t_0}^{t} F(x^n(s), s)ds.$$

We claim that, for some constants C_0, C_1, and all n,

$$\delta^n(t) \equiv ||x^{n+1}(t) - x^n(t)|| \leq C_0 C_1^n \frac{|t - t_0|^n}{n!}.$$

This is true for $n = 0$ if C_0 is chosen big enough, which we assume. Suppose the inequality true up to $n - 1$: Subtracting the equations for x^{n+1} and x^n we obtain

$$\delta^n(t) = || \int_{t_0}^{t} [F(x^n(s), s) - F(x^{n-1}(s), s)]ds||.$$

Since F is C^1 on the compact R', there exists a constant C such that, on R',

$$||F(x, t) - F(y, t)|| \leq C||x - y||.$$

Using the induction hypothesis, we obtain for $t \geq t_0$, say,

$$\delta^n(t) \leq C \int_{t_0}^{t} ||x^n(s) - x^{n-1}(s)||ds \leq CC_0 C_1^{n-1} \int_{t_0}^{t} (s - t_0)^{n-1} \frac{ds}{(n-1)!}$$

$$= CC_0 C_1^{n-1} \frac{(t - t_0)^n}{n!}.$$

This is the desired estimate if $C_1 \geq C$. Hence $||x^{n+1} - x^n||_{L^\infty(J)} \leq C_0(C_1\alpha)^n/n!$, which is the general term of a convergent series. Thus $x^{n+1} - x^n$ is a normally converging sequence in $C^0(J)$, and x^n converges uniformly to some $x \in C^0(J)$, which is solution of the integral equation. $\square$

The uniqueness part of the theorem results from the following stronger result.

Theorem A.2 (Global Uniqueness Theorem). *Let x and y be two solutions of the Cauchy problem on some interval $J \subset I$ containing t_0. Then $x \equiv y$.*

It suffices to prove the theorem for J compact. Then, for some C,

$$||x(t) - y(t)|| \leq C \int_{t_0}^t ||x(s) - y(s)||ds,$$

and the Gronwall Lemma (Lemma 2.16) implies the result. $\square$

We gave the proof of the Cauchy–Lipschitz theorem in details, since the proof of the existence theorem in Section 2.6 is modeled after it.

Using these theorems, it is easy to establish the following result.

Theorem A.3 (Maximal Interval Theorem). *There exists a unique maximal solution $x \in C^1(J)$ of the Cauchy problem defined on an open interval $J =]T_*, T^*[$. If F is defined on $\mathbf{R}^n \times]a, b[$ and $T^* < b$, then*

$$||x(t)|| \to +\infty, \ t \to T^*, \ t < T^*.$$

In this statement, "maximal solution" means that there exists no solution y defined on $K \supset J$ and (strictly) extending x. A simple illustration of this theorem is the *scalar* Cauchy problem

$$x'(t) = F(x(t)), \ x(0) = x_0, \ F \in C^1(\mathbf{R}), \ F > 0.$$

Let $G(x) = \int_0^x \frac{ds}{F(s)}$ be a primitive of $1/F$. For any solution $x \in C^1(I)$,

$$\frac{d}{dt}[G(x(t))] = \frac{1}{F}(x(t)) \times x'(t) = 1,$$

hence $G(x(t)) = G(x_0) + t$, and x will exist as long as $G(x_0) + t$ is in the range of G. Suppose for instance

$$\int_{-\infty}^0 \frac{ds}{F(s)} = \infty, \ \alpha = \int_0^{+\infty} \frac{ds}{F(s)} < \infty.$$

Then G is a strictly increasing function from $-\infty$ to α, and the maximal interval is $]-\infty, T^*[$ with $G(x_0) + T^* = \alpha$, that is $T^* = \int_{x_0}^{+\infty} \frac{ds}{F(s)}$. The other three cases are handled similarly.

It remains for us to understand how the maximal interval depends on the initial value x_0. Though this is a difficult problem, one can easily obtain the following theorem.

Theorem A.4. *Let $\bar{x} \in C^1(]T_*, T^*[)$ be the maximal solution of the Cauchy problem*

$$x'(t) = F(x(t), t), \ x(t_0) = \bar{x}_0.$$

Fix a and b such that $T_* < a < t_0 < b < T^*$. Then there exists $\epsilon > 0$ such that all solutions with initial data $x(t_0) = x_0$ satisfying $\|x_0 - \bar{x}_0\| \leq \epsilon$ are defined on a maximal interval containing $[a, b]$.

For instance, the solution of $x'(t) = x^2(t)$, $x(0) = x_0$ is $x(t) = x_0/(1 - tx_0)$. If we take $\bar{x}_0 = 0$, the solution $\bar{x}$ is global; the solution with data x_0 will be defined on $[-M, M]$ as soon as $|x_0| < 1/M$.

A.1.2   Flows

In the special case when F does not depend on t, we call the system **"autonomous."** It is enough then to consider the Cauchy problem

$$x'(t) = F(x(t)), \ x(0) = x_0.$$

The solution is denoted by $\Phi(t, x_0)$, and called the flow of F. The point of this notation is to emphasize the dependence of the solution on its initial value x_0, and this is very convenient, as we shall see in applications (See Chapters 1–3). By definition, for each x_0, the function $\Phi(t, x_0)$ is defined on the maximal interval $]T_*(x_0), T^*(x_0)[$; hence Φ is defined on $U \subset \mathbf{R} \times \mathbf{R}^n$

$$U = \{(t, x), x \in \Omega, T_*(x) < t < T^*(x)\}.$$

The important result about Φ is the following.

Theorem A.5 (Flow Theorem). *Let Φ be the flow of F. Then U is open and $\Phi \in C^1(U)$.*

We do not prove this theorem (though it can be obtained as an application of the implicit function theorem), but explain why U is open. Let $m_0 = (t_0 > 0, x_0) \in U$: This implies $t_0 + \eta < T^*(x_0)$ for some $\eta > 0$, hence $[0, t_0 + \eta]$ is contained in $]T_*(x_0), T^*(x_0)[$. Using the above theorem about the maximal interval, we obtain that for some $\epsilon > 0$, $T^*(x) > t_0 + \eta$ if $||x - x_0|| \leq \epsilon$. Hence $B(x_0, \epsilon) \times]0, t_0 + \eta[$ is an open set contained in U and containing m_0.

A.1.3 Lower and Upper Fences

Consider a *scalar* equation $x'(t) = F(x(t), t)$.

Definition A.6. *A real function* $y \in C^1([a, b[)$ *is called a lower fence (resp., an upper fence) for the equation* $x'(t) = F(x(t), t)$ *if* $y'(t) \leq F(y(t), t)$ *(resp.,* $y'(t) \geq F(y(t), t)$*).*

The point of this definition lies in the following theorem.

Theorem A.7 (Fence Theorem). *Suppose we are given a solution* $x \in C^1([a, b[)$ *of the scalar equation* $x'(t) = F(x(t), t)$. *If* $y \in C^1([a, b[)$ *is a lower fence (resp., an upper fence) with* $y(a) \leq x(a)$ *(resp.,* $y(a) \geq x(a)$*), then for all* $t \in [a, b[$, $y(t) \leq x(t)$ *(resp.,* $y(t) \geq x(t)$*).*

Since $F \in C^1$, the proof is very simple, and we give it for a lower fence: Let $\delta(t) = x(t) - y(t)$, $\delta(a) \geq 0$; then

$$\delta'(t) \geq F(x(t), t) - F(y(t), t) = \alpha(t)\delta(t),$$

$$\alpha(t) = \int_0^1 (\partial_x F)(sx(t) + (1-s)y(t), t)ds,$$

and the function α is continuous. Setting $A(t) = \int_a^t \alpha(s)ds$ and $z(t) = \delta(t)e^{-A(t)}$, we obtain

$$z'(t) = e^{-A(t)}(\delta'(t) - \alpha(t)\delta(t)) \geq 0, \ z(a) \geq 0.$$

Hence $z(t) \geq 0$ in $[a, b[$, which implies $\delta(t) \geq 0$. $\square$

A.2 Submanifolds

We refer here to the book of M. Spivak [22].

A.2.1 First Definitions

Definition A.8. *A set* $S \subset \mathbf{R}_x^n$ *is a submanifold of dimension* d *if, for all* $x_0 \in S$, *there exists a* C^1-*diffeomorphism* ϕ *from a neighborhood* U *of* x_0

onto a neighborhood V of the origin in $\mathbf{R}_y^n$ such that

$$\phi(S \cap U) = V \cap \Pi_d,$$

where $\Pi_d = \{y \in \mathbf{R}^n, y_{d+1} = \cdots = y_n = 0\}$.

In other words, looking at S through the "glasses" ϕ, we see only a piece of d-plane. Obviously, if we split the coordinates in $\mathbf{R}^n$ as

$$x = (y, z), \ y = (x_1, \ldots, x_d), \ z = (x_{d+1}, \ldots, x_n),$$

the set S defined by $S = \{x, z = f(y)\}$ for some $f \in C^1$ (the graph of f) is a submanifold of dimension d.

Definition A.9. *Suppose $x_0 \in S \subset \mathbf{R}^n$ is a submanifold of dimension d and there exists a curve $x \in C^1(]-\eta, \eta[)$ in $\mathbf{R}^n$ with $x(t) \in S$, $x(0) = x_0$. Then $x'(0)$ is called a tangent vector to S at x_0.*

From these definitions, we obtain easily the following theorem.

Theorem A.10. *The set of all tangent vectors to S at x_0 is a subspace of dimension d denoted by $T_{x_0}S$, called "tangent plane to S at x_0."*

In practice, submanifolds turn out to be defined in two different ways: By a set of equations, or as parametrized surfaces.

A.2.2   Submanifolds Defined by Equations

The simplest case is this:

Theorem A.11. *Let $f \in C^1(\mathbf{R}^n)$ be a real function with $\nabla f \neq 0$. Then*

$$S = \{x \in \mathbf{R}^n, f(x) = 0\}$$

is a submanifold of dimension $n-1$, whose tangent plane T_mS is orthogonal to $\nabla f(m)$.

More generally, let $f_1, \ldots, f_q \in C^1(\mathbf{R}^n)$ be q given real functions:

Theorem A.12. *If all f_i vanish at m and the differentials $D_mf_1, \ldots, D_mf_q$ are independent, the set $S = \{x \in \mathbf{R}^n, f_1(x) = \cdots = f_q(x) = 0\}$ is a submanifold of dimension $n - q$ in a neighborhood of m. The tangent plane T_mS is the intersection of the kernels of the D_mf_i.*

Proof: To see this, let us complete the free system of the q vectors $\nabla f_1(m), \ldots, \nabla f_q(m)$ by vectors $a_{q+1}, \ldots, a_n$ into a basis of $\mathbf{R}^n$. Set now $g_i(x) = a_i \cdot (x - m)$, $i = q + 1, \ldots, n$. Define the map $\psi : \mathbf{R}_x^n \to \mathbf{R}_y^n$ by

$$x \mapsto y = \psi(x) = (f_1(x), \ldots, f_q(x), g_{q+1}(x), \ldots, g_n(x)), \ \psi(m) = 0.$$

The differential $D_m\psi$ is represented by a matrix whose lines form a basis of $\mathbf{R}^n$, hence it is invertible. By the impicit function theorem, ψ is a local diffeomorphism from a neighborhood U of m onto a neighborhood V of 0. The image of $S\cap U$ by ψ is the piece in V of $n-q$ plane $\{y_1 = \cdots = y_q = 0\}$. Hence S is a submanifold of dimension $n-q$.

If $x \in C^1(]-\eta,\eta[)$ is a curve on S, $f_i(x(t)) = 0$ for all i, hence, by differentiation, $x'(0)$ belongs to the kernel of $D_m f_i$. Since this happens for all i, $x'(0)$ belongs to the intersection E of these kernels. Thus, $T_m S$ has dimension $n-d$ and is included in E which also has dimension $n-d$, and this implies $T_m S = E$.   □

The condition about the differentials of the defining functions f_i is easy to understand. Suppose $q = 2$: each equation $f_i = 0$ defines a submanifold S_i of codimension 1, and $S = S_1 \cap S_2$. The condition that ∇f_1 and ∇f_2 be independent just means that S_1 and S_2 are not tangent at m, which is a very reasonable requirement.

A.2.3   Parametrized Surfaces

Let $f : \mathbf{R}^p_u \supset \Omega \to \mathbf{R}^n, f(u) = (f_1(u),\dots,f_n(u))$ be a C^1 function, and set

$$S = \{x \in \mathbf{R}^n, \exists u \in \Omega, \, x = f(u)\}.$$

Intuitively, S, a set of points depending on the p parameters $(u_1,\dots,u_p)$, should be a submanifold of dimension p.

Theorem A.13. *Assume $m_0 = f(u_0)$ and $D_{u_0} f$ injective. Then there exists a neighborhood U of u_0 such that $f(U) \subset S$ is a submanifold of dimension p. The tangent space $T_{m_0}[f(U)]$ is spanned by the vectors $(\partial_1 f(u_0),\dots, \partial_p f(u_0))$.*

The simplest example is a curve $p = 1$, for which the condition of the theorem is just $f'(u_0) \neq 0$, defining the tangent to the curve. In general, consider the $(n \times p)$-matrix representing $D_{u_0} f$: Its columns are the vectors $\partial_i f(u_0)$, which are independent since the differential is injective. Hence there is a $p \times p$ block B, say the first p lines, which is invertible. This block B is the differential at u_0 of the map

$$\phi : \Omega \to \mathbf{R}^p, \ \phi(u) = (f_1(u),\dots,f_p(u)).$$

Let p_0 be the projection of m_0 on the subspace generated by the first p vectors. Since B is invertible, ϕ is a C^1 diffeomorphism from a neighborhood U of u_0 onto a neighborhood V of the projection p_0. Then $f(U)$ is the graph of $f(\phi^{-1})$ over V, hence a submanifold of dimension p.

To visualize $T_{m_0}[f(U)]$, consider the "coordinate curve"

$$u_i \mapsto ((u_0)_1, \ldots, (u_0)_{i-1}, u_i, (u_0)_{i+1}, \ldots, (u_0)_p).$$

This is the parallel to the u_i-axis through u_0. The image of this curve by f is a C^1 curve on S with tangent, by definition, $\partial_i f(u_0)$. Therefore these vectors are tangent vectors and span a subspace of $T_{m_0}[f(U)]$ of dimension p, that is, the whole of the tangent space. $\qquad \square$

A.2.4 Graphs

Let the coordinates in $\mathbf{R}_x^n$ be split as $x = (y, z), y \in \mathbf{R}^p, z \in \mathbf{R}^q, p + q = n$. The subspace $\mathbf{R}_y^p$ is thought of as "horizontal," the subspace $\mathbf{R}_z^q$ as "vertical." Let

$$f : \mathbf{R}_y^p \supset \omega \to \mathbf{R}^{n-p}, \; f(y) = (f_1(y), \ldots, f_{n-p}(y))$$

be a C^1 function, and

$$S = \{x = (y, z) \in \mathbf{R}^n, \; y \in \omega, \; z = f(y)\}.$$

We call S the graph of f over ω. Then the tangent space to S at $m_0 = (y_0, z_0)$ does not contain any vertical vector $(0, V)$. Conversely, we have the following theorem.

Theorem A.14. *Let S be a submanifold of dimension p such that $T_m S$ does not contain any vertical vector. Then S is the graph of some C^1 function in a neighborhood of m.*

Proof: Let S be defined by independent equations $g_1 = \cdots = g_q = 0$, and define

$$g : \mathbf{R}^n \to \mathbf{R}^q, \; g(x) = (g_1(x), \ldots, g_q(x)), \; g(m) = 0.$$

The last $n - p$ columns of the $(n - p \times n)$-matrix representing $D_m g$ form a square block B. The assumption about $T_m S$ means that no (nonzero) vector of the form $(0, V)$ is in the kernel of $D_m g$, and since $D_m g(0, V) = BV$, this means that B is invertible. Now we can use the implicit function theorem at m to solve the equation $g(y, z) = 0$ for z, since $\partial_z g(m) = B$. This yields a C^1 function $f : \mathbf{R}_y^p \to \mathbf{R}_z^q$ defined near the projection p of m, for which

$$S = \{x = (y, z), \; z = f(y)\}.$$

$\qquad \square$

A.2.5 Weaving

Consider, in $\mathbf{R}_x^n$, a p-submanifold Σ containing m $(p < n)$, and a function $F : \mathbf{R}^n \to \mathbf{R}^n$ of class C^1 in a neighborhood of m. The flow of F, defined on an open set U, is denoted by Φ.

Theorem A.15 (Weaving). *Assume that $F(m)$ is not tangent to Σ at m. Then there exists a neighborhood V of $(m, 0)$ in U such that*

$$S = \{x = \Phi(t, y), \ y \in \Sigma, \ (y, t) \in V\}$$

is a $(p + 1) - submanifold$.

Proof: Geometrically, S is the union of the trajectories of the system $x'(t) = F(x(t))$ starting from points of Σ. Let Σ be properly parametrized by $u \in \mathbf{R}^p$, that is, assume that there exists $f : \mathbf{R}_u^p \supset \omega \to \mathbf{R}_x^n$, $0 \in \omega$, $f(0) = m$, such that $\Sigma = f(\omega)$. As in Theorem A.13, assume that the vectors $\partial_{u_i} f(0)$ are independent. In this case, S is naturally parametrized by

$$(t, u) \mapsto \psi(t, u) = \Phi(t, f(u))$$

for (t, u) close to $(0, 0)$. According to Theorem A.13, we need only prove that the vectors $(\partial_t \psi, \partial_{u_1} \psi, \dots, \partial_{u_p} \psi)$ are independent. But

$$\partial_t \psi(0, 0) = F(m), \ \partial_{u_i} \psi(0, 0) = \partial_{u_i} f(0),$$

and since F is not tangent to Σ, these vectors are independent. $\qquad\square$

A.2.6 Stokes Formula

We do not give this formula in full generality, but mention only two very useful special cases.

Formula A.16 (Green-Riemann formula). *In the plane $\mathbf{R}_{x,y}^2$, let D be a compact domain such that its boundary ∂D is piecewise C^1 and can be oriented clockwise. Then for $P, Q \in C^1(D)$,*

$$\int_D (\partial_x Q - \partial_y P) dx dy = \int_{\partial D} P dx + Q dy.$$

The meaning of the integral on the right is

$$\int_a^b [P(x(t), y(t)) x'(t) + Q(x(t), y(t)) y'(t)] dt$$

for a C^1 parametrization $[a, b] \ni t \mapsto (x(t), y(t))$ of ∂D.

In $\mathbf{R}_x^3$, a slightly different formulation is customary.

Formula A.17 (Stokes formula). *Let $D \subset \mathbf{R}_x^3$ be a compact domain with (piecewise) C^1 boundary ∂D. Let*

$$X : D \to \mathbf{R}_x^3, \ X(x) = (X_1(x), X_2(x), X_3(x))$$

be a C^1 function.

Then

$$\int_D [\Sigma \partial_i (X_i(x))] dx = \int_{\partial D} [\Sigma X_i(x) N_i(x)] d\sigma.$$

Here, $N = (N_1, N_2, N_3)$ is the unit outward normal to ∂D, and $d\sigma$ is its surface element. We say that the integral in D of the divergence of X equals the outgoing flux of X through ∂D.

References

[1] Alinhac, S.(1973). Une caractérisation des fonctions harmoniques dans un ouvert borné par des propriétés de moyenne sur certaines boules, *Revue Roumaine Math.* **8**: 1465–1470.

[2] Alinhac, S.(2004). Remarks on energy inequalities for wave and Maxwell equations on a curved background, *Math. Annalen* **329**: 707–722.

[3] Alinhac, S.(2006). On the Morawetz-Keel-Smith-Sogge inequality for the wave equation on a curved background, *Publ. Res. Inst. Math. Sc. Kyoto* **42**: 705–720.

[4] Alinhac, S.(1995). *Blowup for nonlinear hyperbolic equations.* Boston: Birkhäuser.

[5] Alinhac, S., and Gérard, P.(2007). *Pseudo-differential operators and the Nash-Moser theorem.* Rhode Island: American Math. Soc.

[6] Christodoulou, D., and Klainerman, S.(1990). Asymptotic properties of linear field equations in Minkowski space, *Comm. Pure Appl. Math.* **XLIII**: 137–199.

[7] Courant, R. and Friedrichs, K.O.(1948). *Supersonic Flow and Shock Waves.* New York: Interscience Publishing.

[8] Courant, R. and Hilbert, D.(1966). *Methods of Mathematical Physics II.* New York: Interscience Publishing.

[9] Evans, L. C.(1998). *Partial Differential Equations.* Rhode Island: Amer. Math. Soc.

[10] Hörmander, L.(1997). *Lectures on Nonlinear Hyperbolic Differential Equations.* New York: Springer-Verlag.

[11] Hubbard, J. H. and West, B. H.(1991). *Differential Equations: A Dynamical System Approach.* New York: Springer-Verlag.

[12] John, F.(1982).   *Partial Differential Equations.*   New York:
     Springer-Verlag.

[13] Keel, M., Smith, H., and Sogge, C.(2002). Almost global existence
     for some semilinear wave equations, *J. Anal. Math.* **87**: 265–279.

[14] Klainerman, S.(2001).   A commuting vector fields approach to
     Strichartz type inequalities and applications to quasilinear wave equa-
     tions, *Int. Math. Res. Notices* **5**: 221–274.

[15] Lax, P. D.(2006).   *Hyperbolic Partial Differential Equations.*
     New York: Amer. Math. Soc.

[16] Lindblad, H., and Rodnianski, I.(2004).   Global existence for the
     Einstein vacuum equations in wave coordinates, *Preprint.*

[17] Majda, A. J.(1984).   *Compressible Fluid Flow and Systems of
     Conservation Laws in several variables.* New York: Springer-Verlag.

[18] Morawetz, C.(1968).   Time decay for the nonlinear Klein–Gordon
     equation, *Proc. Roy. Soc.* **A306**: 291–296.

[19] Serre, D.(1996). *Systèmes de lois de conservation.* Paris: Diderot Ed.

[20] Shatah, J. and Struwe, M.(1998).   *Geometric Wave Equations.*
     New York: Amer. Math. Soc.

[21] Smoller, J.(1967). *Shock Waves and Reaction–Diffusion Equations.*
     New York: Springer-Verlag.

[22] Spivak, M.(1965). *Calculus on manifolds.* New York: W. A. Benjamin.

[23] Taylor, M. E.(1996).   *Partial Differential Equations.*   New York:
     Springer-Verlag.

[24] Vainberg, B. R.(1989). *Asymptotic Methods in Equations of Math.
     Physics.* New York: Gordon and Breach Publ.

[25] Zuily, C.(2002). *Eléments de distributions et d'équations aux dérivées
     partielles.* Paris: Dunod.

Index